Sure Success in Convergence

Sure Success in Convergence

HARI KISHAN
M.Sc. (Maths), Gold Medalist, Ph.D.
Senior Reader
Department of Mathematics
Kishori Raman Postgraduate College
MATHURA

ATLANTIC®
PUBLISHERS & DISTRIBUTORS

Published by
ATLANTIC®
PUBLISHERS & DISTRIBUTORS
B-2, Vishal Enclave, Opp. Rajouri Garden,
New Delhi-110027
Phones : 25413460, 25429987, 25466842

Sales Office
7/22, Ansari Road, Darya Ganj,
New Delhi-110002
Phones : 23273880, 23275880, 23280451
Fax : 91-11-23285873
web : www.atlanticbooks.com
e-mail : info@atlanticbooks.com

Printed in India
at Nice Printing Press, Delhi

Dedicated to
My Mother
Late Smt. Chandri Devi

Contents

Despite my best efforts, certain errors, misprints, rough-spots, omissions might have crept in the body of the text. I believe that the scope of improvement always exists in a work of this kind. I look forward for the penetrating comments and useful suggestions from esteemed fellow professors and from the students too. If the readers acquaint me of any such points, I shall feel highly grateful to them and incorporate them in the next edition.

198/A, Kachchi Mandi,
Kosi Kalan-281403
(05662) - 233339

Dr. Hari Kishan

Preface

The present book *Sure Success in Convergence* has been designed for the benefit of students of B.Sc. and B.A. courses of Indian universities. This is quite a new subject for the students as they read it for the first time at this stage. The inherent difficulties of the subject are also increased by inadequate treatment given to the subject. Therefore, my focus throughout has been on providing a systematic, logical and interesting treatment to the subject so that even an average student may independently find it accessible to himself.

This book has grown out of my experience in teaching this subject for a period of more than three decades. I have explained all the elementary principles and fundamental concepts rigorously, leaving no scope for illusion or confusion.

The language of the book is simple and style lucid. The examples are numerous, fully solved and well-graded. They have mostly been selected from authoritative textbooks, papers of different universities and competitive examinations. These will help the students in making them familiar with the type of the questions which are usually set at the examinations.

I tender my cordial thanks to my wife, Smt. Hira Mani Agrawal, who has been a source of inspiration to me and has helped me most efficiently by creating and maintaining a suitable family atmosphere for the preparation of this book. I owe much to my son Manish Goyal, Lecturer (Maths.), G.L.A.I.I.M., Mathura, also an author, who has worked as a positive catalyst for me and has assisted me in many ways. I desire to acknowledge my indebtness to my publisher for an excellent production of the book.

1

Convergence and Divergence of Infinite Series

1.1 Series

An expression of the form $u_1 + u_2 + u_3 + \ldots + u_n + \ldots$ in which each term is obtained from its preceding term by some definite law is called a series. This definite law is known as law of formation of the series. $u_1, u_2, u_3, \ldots, u_n, \ldots$ are respectively known as first, second, third, ..., n^{th}, ..., term of the series.

1.2 Finite Series

A series which contains a finite number of terms is called a finite series. Thus,

$$\sum_{r=1}^{n} u_r = u_1 + u_2 + u_3 + \ldots + u_r + \ldots + u_n \text{ is the finite series.}$$

1.3 Infinite Series

A series which contains an infinite number of terms is called an infinite series. In other words, we can say that a series in which every term is followed by another is called an infinite series. Thus,

$$\sum_{n=1}^{\infty} u_n = u_1 + u_2 + u_3 + \ldots + u_n + \ldots \text{ is the infinite series.}$$

The above infinite series may simply be denoted by Σu_n.

The sum of the first n terms of the series is denoted by S_n. Thus,

$$S_n = u_1 + u_2 + u_3 + \ldots + u_n.$$

Note : We shall consider the infinite series only throughout this book, therefore, we are leaving the word infinite here with the understanding that, in what follows, the word series means an infinite series unless otherwise stated.

1.4 Positive Term Series

The series

$$\Sigma u_n = u_1 + u_2 + u_3 + \dots + u_n + \dots$$

is said to be a positive term series if every term of the series is positive. Similarly we can define a negative term series.

1.5 Alternating Series

If the terms of a series are alternately positive and negative beginning with the first term, then, the series is said to be an alternating series. Thus,

$$u_1 - u_2 + u_3 - u_4 + \dots + (-1)^{n+1} u_n + \dots$$

is an alternating series.

1.6 Some Important Limits

1. $$\lim_{n \to \infty} \left(1 + \frac{x}{n}\right)^n = e^x$$

Proof.

$$\lim_{n \to \infty} \left(1 + \frac{x}{n}\right)^n = e^x$$

$$= \lim_{n \to \infty} \left[1 + n.\frac{x}{n} + \frac{n(n-1)}{2!}\frac{x^2}{n^2} + \frac{n(n-1)(n-2)}{3!}\frac{x^3}{n^3} + \dots\dots\right]$$

$$\lim_{n \to \infty} \left[1 + x + \frac{\left(1 - \frac{1}{n}\right)}{2!}x^2 + \frac{\left(1 - \frac{1}{n}\right)\left(1 - \frac{2}{n}\right)}{3!}x^3 + \dots\dots\right]$$

$$= 1 + x + \frac{x^2}{2!} + \frac{x^3}{3!} + \dots\dots$$

$$= e^x.$$

It can be easily deduced from above that

$$\lim_{n \to \infty} \left(1+\frac{1}{n}\right)^n = e.$$

2. $\lim_{n\to\infty} \dfrac{\log n}{n} = 0.$

Proof. Put $\log n = t$, so that we have, $n = e^t$, then

$$\lim_{n\to\infty} \frac{\log n}{n} = \lim_{t \to \infty} \frac{t}{e^t}$$

$[\because$ if $n\to\infty$, then $t\to\infty$ from $t = \log n]$

$$= \lim_{t \to \infty} \frac{t}{1+t+\dfrac{t^2}{2!}+\dfrac{t^3}{3!}+\ldots}$$

$$= \lim_{t \to \infty} \frac{t}{\dfrac{1}{t}+1+\dfrac{t}{2!}+\dfrac{t^2}{3!}+\ldots}$$

[dividing the numerator and denominator by t]

$$= \frac{1}{\infty}$$

$$= 0.$$

3. $\lim_{n\to\infty} nx^n = 0$, **when** $0 < x < 1$.

Proof. If $x < 1$,

then $\dfrac{1}{x} > 1$

and therefore

$$\lim_{n\to\infty} \left(\frac{1}{x}\right)^n \to \infty.$$

Let, $\left(\dfrac{1}{x}\right)^n = t$, so that $x^n = \dfrac{1}{t}$.

Taking log on both sides, we get

$$n\log\left(\frac{1}{x}\right)=\log t$$

or, $$n=\frac{\log t}{\log\left(\frac{1}{x}\right)}$$

or, $$nx^n=\frac{\log t}{\log\left(\frac{1}{x}\right)}\cdot x^n$$

or, $$nx^n=\frac{\log t}{\log\left(\frac{1}{x}\right)}\cdot\frac{1}{t} \qquad \left[\because\ x^n=\frac{1}{t}\right]$$

or, $$nx^n=\frac{\log t}{t}\cdot\frac{1}{\log\left(\frac{1}{x}\right)}$$

4. $$\underset{n\to\infty}{\text{lt.}}\frac{A_1n^p+A_2n^{p-1}+A_3n^{p-2}+\ldots+A_{p+1}}{B_1n^q+B_2n^{q-1}+B_3n^{q-2}+\ldots+B_{q+1}}=0 \text{ or } \infty$$

Proof.

$$\underset{n\to\infty}{\text{lt.}}\frac{A_1n^p+A_2n^{p-1}+A_3n^{p-2}+\ldots+A_{p+1}}{B_1n^q+B_2n^{q-1}+B_3n^{q-2}+\ldots+B_{q+1}}$$

0 if $p<q$

$\frac{A_1}{B_1}$ if $p=q$

∞ if A_1 and B_1 are of the same signs and $p>q$

$-\infty$ if A_1 and B_1 are of the opposite signs and $p>q$

$$=\underset{n\to\infty}{\text{lt.}}\frac{n^p\left(A_1+\frac{A_2}{n}+\frac{A_3}{n^2}+\ldots+\frac{A_{p+1}}{n^p}\right)}{n^q\left(B_1+\frac{B_2}{n}+\frac{B_3}{n^2}+\ldots+\frac{B_{q+1}}{n^q}\right)}$$

$$=\underset{n\to\infty}{\text{lt.}}\frac{n^{p-q}\left(A_1+A_2.\frac{1}{n}+A_3\frac{1}{n^2}+\ldots+\frac{A_{p+1}}{n^p}\right)}{\left(B_1+B_2.\frac{1}{n}+B_3.\frac{1}{n^2}+\ldots+\frac{B_{q+1}}{n^q}\right)}$$

$= 0$ if $p < q, \frac{A_1}{B_1}$ if $p = q$, ∞ if A_1 and B_1 are of the same signs while $p > q$ and $-\infty$ if A_1 and B_1 are of the opposite signs while $p > q$.

5. $\underset{n \to \infty}{\textbf{lt.}} \; x^n = 0 \quad$ **(if x < 1)**

$\qquad\qquad = \infty \quad$ (if x > 1)

Proof.

Take lt. as $n \to \infty$ on both sides, we get

$$\underset{n \to \infty}{\text{lt.}} \; nx^n = \underset{t \to \infty}{\text{lt.}} \; \frac{\log t}{t} \cdot \frac{1}{\log\left(\frac{1}{x}\right)}$$

$$= 0$$

because from (2) above $\underset{t \to \infty}{\text{lt.}} \; \frac{\log t}{t} = 0$

and $\frac{1}{\log\left(\frac{1}{x}\right)}$ is a finite quantity.

Thus we prove the limit.

1.7 Convergence and Divergence of a Series: Definition

Let $\Sigma u_n = u_1 + u_2 + u_3 + \ldots u_n + \ldots$

be an infinite series. Then,

$$S_n = u_1 + u_2 + u_3 + \ldots + u_n$$

and we define the convergence and divergence of the given series as follows :

(i) If $\underset{n \to \infty}{\text{lt.}} \; S_n = S$, where S is a finite and unique quantity, then $\Sigma\, u_n$ is convergent. We say the series itself converges to S. S is known as sum of the series.

(ii) If $\underset{n \to \infty}{\text{lt.}} \; S_n = +\infty$ or $-\infty$, then $\Sigma\, u_n$ is divergent. We say that the series itself diverges to $+\infty$ or $-\infty$, as the case may be.

(iii) If $\underset{n \to \infty}{\text{lt.}} \; S_n$ is neither a finite unique quantity, nor $-\infty$ or $+\infty$, then the series $\Sigma\, u_n$ is said to be oscillatory. Some mathematicians call it as periodic convergent series. A series can oscillate finitely or infinitely.

(a) The series Σu_n is said to oscillate finitely if $\underset{n \to \infty}{\text{lt.}} S_n$ is a finite but not a unique quantity, i.e. $\underset{n \to \infty}{\text{lt.}} S_n$ fluctuates between finite limits.

(b) The series Σu_n is said to oscillate infinitely if $\underset{n \to \infty}{\text{lt.}}$ S_n fluctuates between $-\infty$ and $+\infty$.

Examples

Consider the series

$$1+\frac{1}{2}+\frac{1}{2^2}+\frac{1}{2^3}+\dots$$

Here,

$$S_n = \frac{1\left(1-\frac{1}{2^n}\right)}{1-\frac{1}{2}}$$

or, $S_n = 2\left(1-\frac{1}{2^n}\right)$

Clearly,

$\underset{n \to \infty}{\text{lt.}} S_n = 2$ which is a finite unique quantity, hence the series is convergent.

Again, consider the series

$$1+2+3+4+5+\dots$$

Here,

$$S_n = \frac{n(n+1)}{2}$$

Clearly,

$$\underset{n \to \infty}{\text{lt.}} S_n = \infty$$

Hence, the series is divergent.

Further, consider the series

$$-1-2-3-4-\ldots$$

Here,

$$S_n = -\frac{n(n+1)}{2}$$

Clearly,

$$\underset{n\to\infty}{\text{lt.}} S_n = -\infty$$

Hence, the series is divergent.

Now, consider the series

$$1-1+1-1+1-1+\ldots$$

Here,

$$S_n = 0 \text{ if } n \text{ is even and}$$
$$= 1 \text{ if } n \text{ is odd.}$$

Thus S_n oscillates (fluctuates) between 0 and 1 which are the finite limits. So the series oscillates finitely.

In the last, consider the series

$$1-2+3-4+5-6+\ldots$$

Here, $S_1 = 1$, $S_2 = -1$

$S_3 = 2$, $S_4 = -2$

$S_5 = 3$, $S_6 = -3$

and so on.

Thus, $S_n = -\frac{n}{2}$ if n is even

and $S_n = \left(\frac{n+1}{2}\right)$ if n is odd

Hence $\underset{n\to\infty}{\text{lt.}} S_n = -\infty$ if n is even

and $\underset{n\to\infty}{\text{lt.}} S_n = +\infty$ if n is odd.

Therefore, the series oscillates infinitely between $-\infty$ and $+\infty$.

Note 1 : If all the terms of a series are of the same sign (either positive or negative), then this series can either be convergent or divergent and never an oscillatory one.

Note 2 : Now onward, we are to study only convergent or divergent series.

Note 3 : By "lt." we shall mean $\underset{n\to\infty}{\text{lt.}}$.

1.8 Geometric Series

An infinite geometric series with common ratio r is

(i) convergent if $-1 < r < 1$ i.e. $|r| < 1$

(ii) divergent if $r > 1$

(iii) finitely oscillating if $r = -1$

(iv) infinitely oscillating if $r < -1$

Consider the infinite geometric series

$$1 + r + r^2 + r^3 + \ldots + r^{n-1} + r^n + \ldots$$

whose common ratio is r.

Case I. When $-1 < r < 1$ i.e. $|r| < 1$.

We know that, in this case as r is numerically less than 1,

$$\therefore \quad S_n = \frac{1-r^n}{1-r}$$

Take lt. as $n \to \infty$ on both sides, we have

$$\underset{n\to\infty}{\text{lt.}} S_n = \underset{n\to\infty}{\text{lt.}} \frac{1-r^n}{1-r}$$

$$= \frac{1}{1-r} \qquad \left[\begin{array}{l}\text{as } r^n \to 0 \text{ when } n \to \infty, \\ |r| \text{ being less than } 1\end{array}\right]$$

which is a finite unique quantity, hence the series is convergent by definition of convergence of a series.

Case II. When $r > 1$

In this case, as $r > 1$,

$$\therefore \quad S_n = \frac{r^n - 1}{r-1}$$

$$\therefore \lim_{n\to\infty} S_n = \lim_{n\to\infty} \frac{r^n - 1}{r-1} = \infty \qquad \left[\because r^n \to \infty \text{ as } n \to \infty, \; r \text{ being greater than } 1\right]$$

Hence the series is divergent by definition of divergence of a series.

Also, when $r = 1$, the series becomes,

$$1 + 1 + 1 + 1 + 1 + 1 + \ldots$$

Thus, $S_n = n$ which tends to ∞ as $n \to \infty$, hence again the series is divergent.

Case III. When $r = -1$, then the series becomes $1 - 1 + 1 - 1 + 1 - 1 + \ldots$

As already proved, we see that,

$$\lim_{n\to\infty} S_n = 1 \text{ if } n \text{ is odd,}$$
$$= 0 \text{ if } n \text{ is even.}$$

Here the limits are finite but not unique.

Hence the series is finitely oscillating series.

Case IV. When $r < -1$.

If $r < -1$,

then $-r > 1$

Let $x = -r$

then $x > 1$

and therefore

$$x^n \to \infty \text{ a } n \to \infty$$

Now,

$$S_n = 1 + r + r^2 + \ldots + r^{n-1}$$

$$= \frac{1 - r^n}{1 - r}$$

$$= \frac{1 - (-x)^n}{1 - (-x)}$$

$$= \frac{1 - (-x)^n}{1 + x}$$

$$= \frac{1+x^n}{1+x} \text{ or } \frac{1-x^n}{1+x} (\text{where } x > 1)$$

according as n is odd or even.

$$\therefore \underset{n \to \infty}{\text{lt.}} S_n = \frac{1+\infty}{1+x} \text{ or } \frac{1-\infty}{1+x}$$

$$= +\infty \text{ or } -\infty$$

Hence the series is infinitely oscillating.

1.9 It is clear from the treatment given above that in order to decide whether a given series is convergent or divergent, we shall have to first determine S_n and then its limit as $n \to \infty$. The value of limit will let us know about the convergency and divergency of the series. But there are many series in which the determination of S_n is not so easy and rather impossible too. Thus the above treatment fails in case of such series with reference to judgment of their convergence and divergence. Hence some new methods (tests) have been searched out to overcome this problem of deciding the convergence or divergence of a series without actually finding out the sum of its first n terms. Before we discuss those tests, it will be quite appropriate and timely to be acquainted with some fundamental properties of the series related to their convergence or divergence.

1.10 Fundamental Properties

(i) *The nature of a series remains unaltered if the signs of all the terms are altogether changed.*

Let $\Sigma u_n = u_1 + u_2 + u_3 + ... + u_n + ...$

be the given infinite series.

Then,

$$S_n = u_1 + u_2 + u_3 + ... + u_n.$$

If the signs of all the terms of the given series are altogether changed, then, the series will become,

$$-u_1 - u_2 - u_3 - u_4 - ... - u_n - ...$$

Let the sum of the first n terms of this new series be S_n'.

Then,

$$S_n' = -u_1 - u_2 - u_3 - ... - u_n.$$

Clearly, $S'_n = -S_n$

$$\therefore \lim_{n \to \infty} S'_n = -\lim_{n \to \infty} S_n.$$

It is quite clear that if $\lim_{n \to \infty} S_n$ is finite and definite then $\lim_{n \to \infty} S'_n$ will also be finite and definite. i.e. if the given series is convergent, then the new series will also be convergent.

Further, if $\lim_{n \to \infty} S_n = +\infty \text{ or} -\infty$

then $\lim_{n \to \infty} S'_n = +\infty \text{ or} -\infty$

i.e. if the given series is divergent, then the new series is also divergent.

In the last, if $\lim_{n \to \infty} S_n$ oscillates,

then $\lim_{n \to \infty} S'_n$ also oscillates

i.e. if the given series is oscillatory, then the new series will also be oscillatory.

(ii) *The convergence or divergence of a series remains uneffected if a finite number of terms are added or neglected. (omitted or left)*

It can be easily understood that the sum of a finite number of terms is finite and definite. Hence if we add or neglect them, then the sum of the series i.e. $\lim_{n \to \infty} S_n$ will get changed merely by a finite determinate quantity.

Hence if $\lim_{n \to \infty} S_n$ is finite then it will remain finite and if $\lim_{n \to \infty} S_n$ is infinite then it will remain infinite even when a finite number of terms are added or omitted. Thus the convergence or divergence of the given series will remain unaffected.

This property is very commonly used while solving the questions.

(iii) *If each term of a given series is multiplied or divided by some fixed quantity other than zero, then the new series so obtained will remain convergent or divergent according as it was originally convergent or divergent.*

Let the given series be

$$\Sigma u_n = u_1 + u_2 + u_3 + \dots + u_n + \dots$$

and $$S_n = u_1 + u_2 + u_3 + \dots + u_n$$

Let the series be multiplied by some fixed quantity λ, where $\lambda \neq 0$. Then the series will become,

$$\lambda u_1 + \lambda u_2 + \lambda u_3 + \dots + \lambda u_n + \dots$$

Let S'_n be the sum of first n terms of this new series then,

$$S'_n = \lambda u_1 + \lambda u_2 + \lambda u_3 + \dots + \lambda u_n$$
$$= \lambda(u_1 + u_2 + u_3 + \dots + u_n)$$
$$= \lambda S_n$$

$$\therefore \quad \underset{n \to \infty}{\text{lt.}} S'_n = \underset{n \to \infty}{\text{lt.}} \lambda S_n$$

$$= \lambda \underset{n \to \infty}{\text{lt.}} S_n.$$

It is quite obvious from above that

$\underset{n \to \infty}{\text{lt.}} S'_n$ is finite and definite, $+\infty$ or $-\infty$, oscillates according

as $\underset{n \to \infty}{\text{lt.}} S_n$ is finite and definite, $+\infty$ or $-\infty$, oscillates.

Hence, the nature of the new series will remain the same as that of the original series.

Again, let $\lambda = \frac{1}{m}$ (m is finite and $m \neq 0$)

then the series becomes

$$\frac{1}{m}u_1 + \frac{1}{m}u_2 + \frac{1}{m}u_3 + \dots + \frac{1}{m}u_n + \dots$$

With the same reasoning as discussed above we can prove that the nature of the new series in this case even will remain the same as that of the original series.

(iv) *If two infinite series are given, then the series formed by their sum will be :*

(A) *convergent if both the given series are convergent.*

(B) *divergent if any one of the given series is divergent.*

Let $\Sigma u_n = u_1 + u_2 + u_3 + ... + u_n + ...$

and $\Sigma v_n = v_1 + v_2 + v_3 + ... + v_n + ...$

be the given series.

The series formed by their sum is

$$(u_1 + v_1) + (u_2 + v_2) + (u_3 + v_3) + ... + (u_n + v_n) + ...$$

Let S_n be the sum of first n terms of the first series, then,

$$S_n = u_1 + u_2 + u_3 + ... + u_n.$$

Similarly let S_n' be the sum of first n terms of the second series, then,

$$S_n' = v_1 + v_2 + v_3 + ... + v_n.$$

Let S_n'' be the sum of first n terms of the new series formed by their sum, then,

$$\begin{aligned} S_n'' &= (u_1 + v_1) + (u_2 + v_2) + (u_3 + v_3) + ... + (u_n + v_n) \\ &= (u_1 + u_2 + u_3 + ... + u_n) + (v_1 + v_2 + v_3 + ... + v_n) \\ &= S_n + S_n'. \end{aligned}$$

$$\therefore \quad \underset{n \to \infty}{\text{lt.}} S_n'' = \underset{n \to \infty}{\text{lt.}} S_n + \underset{n \to \infty}{\text{lt.}} S_n'.$$

Now if both of the given series are convergent, then,

$$\underset{n \to \infty}{\text{lt.}} S_n \text{ and } \underset{n \to \infty}{\text{lt.}} S_n'$$

will both be finite and definite and consequently

$$\underset{n \to \infty}{\text{lt.}} S_n''$$

will be finite and definite, hence the new series is convergent.

Again if any one of the given series (say first series) is divergent, then, $\underset{n \to \infty}{\text{lt.}} S_n$ will be infinite which when added to a finite and definite quantity $\left(\underset{n \to \infty}{\text{lt.}} S_n' \right)$ will result into an infinite quantity, hence the new series will be divergent.

(v) *The series less than a convergent series is convergent, and greater than a divergent series is divergent.*

(vi) *For a series* Σu_n *to be convergent, it is necessary but not sufficient that*

$$\lim_{n \to \infty} u_n = 0$$

or in other words, for every convergent series Σu_n,

$$\lim_{n \to \infty} u_n = 0$$

Proof. The condition is necessary.

Let the given series

$$\Sigma u_n = u_1 + u_2 + u_3 + \dots + u_n + \dots$$

be convergent.

Let S_n be the sum of the first n terms of this series, then,

$$S_n = u_1 + u_2 + u_3 + \dots + u_{n-1} + u_n.$$

Replacing n by $(n - 1)$ we can get the sum of first $(n - 1)$ terms of the series S_{n-1} as,

$$S_{n-1} = u_1 + u_2 + u_3 + \dots + u_{n-1}.$$

Clearly,

$$u_n = S_n - S_{n-1}$$

$$\therefore \quad \lim_{n \to \infty} u_n = \lim_{n \to \infty} S_n - \lim_{n \to \infty} S_{n-1}.$$

Now since the series Σu_n is convergent, therefore,

$$\lim_{n \to \infty} S_n \text{ and } \lim_{n \to \infty} S_{n-1}$$

both will tend to the same finite and definite limit S.

$$\therefore \quad \lim_{n \to \infty} u_n = S - S$$

$$= 0.$$

The condition is not sufficient.

We are to prove that if $\lim_{n \to \infty} u_n = 0$, then the series Σu_n is not necessarily convergent.

For this, let us consider the series,

$$\Sigma\frac{1}{\sqrt{n}} = 1+\frac{1}{\sqrt{2}}+\frac{1}{\sqrt{3}}+\ldots+\frac{1}{\sqrt{n}}+\ldots.$$

Here, $u_n = \frac{1}{\sqrt{n}}$

$$\therefore \quad \underset{n\to\infty}{\text{lt.}}\ u_n = \underset{n\to\infty}{\text{lt.}}\ \frac{1}{\sqrt{n}} = 0$$

But this series is divergent as has been proved below :

$$S_n = 1+\frac{1}{\sqrt{2}}+\frac{1}{\sqrt{3}}+\ldots+\frac{1}{\sqrt{n}}$$

$$> \frac{1}{\sqrt{n}}+\frac{1}{\sqrt{n}}+\frac{1}{\sqrt{n}}+\ldots+\frac{1}{\sqrt{n}}$$

$$> \frac{n}{\sqrt{n}}$$

$$> \sqrt{n}$$

$$\therefore \quad \underset{n\to\infty}{\text{lt.}}\ S_n > \underset{n\to\infty}{\text{lt.}}\ \sqrt{n}$$

$$> \infty.$$

Hence, the series is divergent.

(vii) *If a series Σu_n converges to a fixed finite quantity S, then the series obtained by grouping the terms in brackets without affecting the order of the terms also converges to S.*

Let $\Sigma u_n = u_1 + u_2 + u_3 + \ldots + u_n + \ldots,$

let us form a new series Σv_n, by grouping the terms of Σu_n in brackets, taking m terms within each pair of brackets and without altering the order of the terms in the given series, as follows :

$$\Sigma v_n = v_1 + v_2 + v_3 + \ldots$$

$$= (u_1 + u_2 + \ldots + u_m)$$

$$+ (u_{m+1} + u_{m+2} + \ldots + u_{2m})$$

$$+ (u_{2m+1} + u_{2m+2} + ... + u_{3m}) +$$

....

$$= \Sigma u_n \text{ actually.}$$

The new series is actually the old series, hence the new series will remain convergent if the original series is convergent.

The converse of this theorem is not true.

Let us consider a series

$$1 - 1 + 1 - 1 + 1 - 1 + ...$$

We know that this is not a convergent series.

Let us now group the terms of this series as follows :

$$(1 - 1) + (1 - 1) + (1 - 1) + ...$$

or, $$0 + 0 + 0 + ...$$

which is convergent.

1.11 A Test for Divergence of a Series

Statement. If all the terms of an infinite series are positive and each term is greater than some fixed finite quantity however small, then the series will be divergent.

Proof. Let $\Sigma u_n = u_1 + u_2 + u_3 + ... + u_n + ...$

be the given series. Let S_n denotes the sum of first n terms of the series, then,

$$S_n = u_1 + u_2 + u_3 + ... + u_n,$$

let each term of the series be greater than some finite quantity say K. Then we have,

$$S_n > K + K + K + ... + K$$

$$> nK$$

Hence,

$$S_n \to \infty \text{ as } n \to \infty$$

Hence, the series Σu_n is divergent.

1.12 Leibnitz's Test

or

Alternating Series Test

Statement. An infinite series in which, the terms are alternately positive and negative.

is convergent if each term is numerically less than the preceding term and if $\underset{n \to \infty}{\text{lt.}} u_n = 0.$

or

The infinite series $\sum_{n=1}^{\infty} (-1)^{n+1} u_n$ is convergent if $u_n > 0$, $u_n < u_{n-1}$ and $\underset{n \to \infty}{\text{lt.}} u_n = 0.$

Let the series be

$$u_1 - u_2 + u_3 - u_4 + u_5 - u_6 + \ldots$$

....

where $u_1 > u_2 > u_3 > u_4 > u_5 > u_6 > \ldots$

Let S_{2n} and S_{2n+1} represent the sum of first $2n$ and first $(2n + 1)$ terms of the series respectively.

Then,

$$S_{2n} = (u_1 - u_2) + (u_3 - u_4) + (u_5 - u_6) + \ldots + (u_{2n-1} - u_{2n})$$

$$= \text{a positive quantity as } u_1 > u_2 > u_3 > \ldots \text{ i.e.} > 0 \quad \ldots (1)$$

and

$$S_{2n+1} = u_1 - [(u_2 - u_3) + (u_4 - u_5) + \ldots + (u_{2n} - u_{2n+1})]$$

$$= u_1 - \text{a positive quantity}$$

$$(\because u_2 > u_3 > u_4 \ldots)$$

$$< u_1 \text{ as } S_{2n+1} \text{ cannot be negative} \quad \ldots (2)$$

Also,

$$S_{2n+1} = S_{2n} + u_{2n+1}$$

$$\therefore \underset{n \to \infty}{\text{lt.}} S_{2n+1} = \underset{n \to \infty}{\text{lt.}} S_{2n} + \underset{n \to \infty}{\text{lt.}} u_{2n+1}$$

$$= \underset{n \to \infty}{\text{lt.}} S_{2n} + 0$$

$$\left(\because \underset{n \to \infty}{\text{lt.}} u_n = 0\right)$$

$$\therefore \underset{n \to \infty}{\text{lt.}} S_{2n+1} = \underset{n \to \infty}{\text{lt.}} S_{2n}$$

i.e. when $n \to \infty$, the sum of an odd number of terms of the series is same as sum of and even number of terms of the series.

Hence the series cannot be oscillating.

Now, from (1) and (2) it is clear that the above common limit must lie between 0 and u_1 and since u_1 is finite and definite, hence the series must be convergent.

Illustrative Examples

Ex. 1. *Examine the series*

$$2+\frac{3}{2}+\frac{4}{3}+\frac{5}{4}+\ldots$$

Sol. We observe that in the given series, each term is positive and greater than 1 which is a finite quantity, hence the series is divergent.

Ex. 2. *Test the series*

$$\sqrt{\frac{1}{2}}+\sqrt{\frac{2}{3}}+\sqrt{\frac{3}{4}}+\sqrt{\frac{4}{5}}+\ldots$$

Sol. If we leave the first term, the series becomes

$$\sqrt{\frac{2}{3}}+\sqrt{\frac{3}{4}}+\sqrt{\frac{4}{5}}+\ldots$$

It is to be noted that the nature of the series does not get affected by doing this.

Now in this series we observe that

(1) all the terms are positive

(2) each term is greater than $\sqrt{\frac{1}{2}}$, which is a finite quantity.

Hence the series is divergent.

Ex. 3. *Test whether the series*

$$\frac{1}{1+2^{-1}}+\frac{2}{1+2^{-2}}+\frac{3}{1+2^{-3}}+\ldots$$

is convergent or divergent.

Sol. The given series may be written as

$$\frac{1}{1+\frac{1}{2}}+\frac{2}{1+\frac{1}{2^2}}+\frac{3}{1+\frac{1}{2^3}}+\ldots$$

i.e. $\frac{2}{3}+\frac{8}{5}+\frac{24}{9}+...$

We observe here that

(1) all the terms are positive

(2) each term is greater than $\frac{1}{2}$ which is a finite quantity.

Hence the given series is divergent.

Ex. 4. *Show that the series*

$$\frac{1}{2}+\frac{1+a}{2+a}+\frac{1+2a}{2+2a}+...+\frac{1+(n-1)a}{2+(n-1)a}+...(a>0) \text{ is divergent.}$$

Sol. $(n+1)^{\text{th}}$ term of this series

$$=u_{n+1}$$

$$=\frac{1+na}{2+na}$$

Now,

$$\frac{1+na}{2+na}-\frac{1}{2}$$

$$=\frac{2+2na-2-na}{2(2+na)}$$

$$=\frac{na}{2(2+na)}$$

which is positive for all integral values of n.

Thus each term of the series is greater than $\frac{1}{2}$ (leaving the first term).

Also all the terms of the series are positive as $a > 0$,

Hence the given series is divergent.

Ex. 5. *Test the convergency or divergency of the series*

$$\frac{a}{b}+\frac{a+x}{b+x}+\frac{a+2x}{b+2x}+\frac{a+3x}{b+3x}+...$$

where a and b are finite and positive, $b > a$ and $x > 0$.

Sol. General term of the series

$$= u_{n+1}$$

$$= \frac{a+nx}{b+nx}$$

Now, $\frac{a+nx}{b+nx} - \frac{a}{b}$

$$= \frac{ab+nbx-ab-anx}{b(b+nx)}$$

$$= \frac{n(b-a)x}{b(b+nx)}$$

which is > 0 (as $b > a$, $x > 0$) for all value of n.

Thus each term is greater than $\frac{a}{b}$ (leaving the first term).

Also all the terms of the series are positive as a and b are positive and $x > 0$.

Hence the given series is divergent.

Ex. 6. *Prove that the series*

$$1 - \frac{1}{2} + \frac{1}{3} - \frac{1}{4} + \ldots$$

is convergent.

Sol. We observe here that

(1) the terms of the series are alternately positive and negative.

(2) each term is numerically less than its preceding term, as

$$1 > \frac{1}{2} > \frac{1}{3} > \frac{1}{4} > \ldots$$

(3) $u_n = \frac{1}{n}$

$\therefore \lim_{n \to \infty} u_n = 0$

Thus all the three conditions of Leibnitz's test are being satisfied by the series.

Hence by Leibnitz's test, the given series is convergent.

Ex. 7. *Test the convergence of the series*

$$1-\frac{1}{3}+\frac{1}{5}-\frac{1}{7}+...$$

Sol. We observe here that

(1) the terms are alternately +ve and –ve

(2) each term is numerically less than its preceding term, as

$$1>\frac{1}{3}>\frac{1}{5}>\frac{1}{7}>...$$

(3) $\underset{n\to\infty}{\text{lt.}}\; u_n$

$$= \underset{n\to\infty}{\text{lt.}}\; \frac{1}{2n-1}$$

$$=\frac{1}{2\times\infty-1}$$

$$=\frac{1}{\infty}$$

$$=0$$

Thus all the three conditions of Leibnitz's test are being satisfied by the series.

Hence by Leibnitz's test, the given series is convergent.

Ex. 8. *Examine the convergence of the series*

$$1-\frac{1}{2}+\frac{1}{4}-\frac{1}{8}+\frac{1}{16}-...$$

Sol. The given series can be written as

$$\frac{1}{2^0}-\frac{1}{2^1}+\frac{1}{2^2}-\frac{1}{2^3}+\frac{1}{2^4}-...$$

We observe in the given series that

(1) the terms are alternately +ve and –ve

(2) each term is numerically less than its preceding term, as

$$1>\frac{1}{2}>\frac{1}{2^2}>\frac{1}{2^3}>\frac{1}{2^4}>...$$

(3) $$u_n = \frac{1}{2^{n-1}}$$

$$\therefore \quad \underset{n\to\infty}{\text{lt.}} u_n = \frac{1}{2^{\infty-1}} = \frac{1}{2^{\infty}} = \frac{1}{\infty} = 0$$

Thus all the three conditions of Leibnitz's test are satisfied by the series.

Hence by Leibnitz's test, the given series is convergent.

Ex. 9. *Show that the series*

$$1 - \frac{1}{\sqrt{2}} + \frac{1}{\sqrt{3}} - \frac{1}{\sqrt{4}} + \ldots$$

is convergent.

Sol. We see that

(1) the terms of the series are alternately +ve and –ve

(2) each term is numerically less than its preceding term, as

$$1 > \frac{1}{\sqrt{2}} > \frac{1}{\sqrt{3}} > \frac{1}{\sqrt{4}} > \ldots$$

(3) $\underset{n\to\infty}{\text{lt.}} u_n$

$$= \underset{n\to\infty}{\text{lt.}} \frac{1}{\sqrt{n}}$$

$$= 0.$$

Thus all the three conditions of Leibnitz's test are satisfied by the series. Hence by Leibnitz's test, the given series is convergent.

Ex. 10. *Test the series*

$$\frac{1}{\sqrt{2}-1} - \frac{1}{\sqrt{3}-1} + \frac{1}{\sqrt{4}-1} - \frac{1}{\sqrt{5}-1} + \ldots$$

Sol. We observe here that

(1) the terms of the series are alternately +ve and –ve

(2) each term is numerically less than its preceding term, as

$$\frac{1}{\sqrt{2}-1} > \frac{1}{\sqrt{3}-1} > \frac{1}{\sqrt{4}-1} > \frac{1}{\sqrt{5}-1} > \ldots$$

(3) $\lim_{n \to \infty} u_n$

$$= \lim_{n \to \infty} \frac{1}{\sqrt{n+1}-1}$$

$= 0$

Thus all the three conditions of Leibnitz's test are satisfied by the series. Hence by Leibnitz's test, the given series is convergent.

Ex. 11. *Test the series*

$$1-\frac{1}{2\sqrt{2}}+\frac{1}{3\sqrt{3}}-\frac{1}{4\sqrt{4}}+\ldots$$

Sol. We observe that in the given series

(1) the terms are alternately +ve and –ve

(2) each term is numerically less than is preceding term, as

$$1>\frac{1}{2\sqrt{2}}>\frac{1}{3\sqrt{3}}>\frac{1}{4\sqrt{4}}>\ldots..$$

(3) $\lim_{n \to \infty} u_n$

$$= \lim_{n \to \infty} \frac{1}{n\sqrt{n}}$$

$= 0.$

Thus all the three conditions of Leibnitz's test are satisfied by the series. Hence by Leibnitz's test, the given series is convergent.

Ex. 12. *Test the series*

$$1-\frac{1}{2^2}+\frac{1}{3^2}-\frac{1}{4^2}+\ldots$$

Sol. We observe that

(1) the terms of the series are alternately +ve and –ve

(2) each term is numerically less than its preceding term, as

$$1>\frac{1}{2^2}>\frac{1}{3^2}>\frac{1}{4^2}>\ldots$$

(3) $\lim_{n\to\infty} u_n$

$$= \lim_{n\to\infty} \frac{1}{n^2}$$

$$= 0$$

Thus all the three conditions of Leibnitz's test are satisfied by the series. Hence by Leibnitz's test, the given series is convergent.

Ex. 13. *Test the series*

$$\frac{2}{1^3}-\frac{3}{2^3}+\frac{4}{3^3}-\frac{5}{4^3}+\ldots$$

Sol. We observe in this series that

(1) the terms are alternately +ve and –ve

(2) each term is numerically less than its preceding term, as

$$\frac{2}{1^3}>\frac{3}{2^3}>\frac{4}{3^3}>\frac{5}{4^3}>\ldots$$

(3) $\lim_{n\to\infty} u_n$

$$= \lim_{n\to\infty} \frac{n+1}{n^3}$$

$$= \lim_{n\to\infty} \frac{1}{n^2}+\frac{1}{n^3}$$

$$= 0.$$

Thus all the three conditions of Leibnitz's test are satisfied by the series. Hence by Leibnitz's test, the given series is convergent.

Ex. 14. *Show that the series*

$$\log\frac{2}{1}-\log\frac{3}{2}+\log\frac{4}{3}-\log\frac{5}{4}+\ldots$$

is convergent.

Sol. We observe in the given series that

(1) the terms are alternately +ve and –ve

(2) each term is numerically less than its preceding term, as

$$\log\frac{2}{1}>\log\frac{3}{2}>\log\frac{4}{3}>\log\frac{5}{4}>\ldots$$

(3) $\lim_{n \to \infty} u_n$

$$= \lim_{n \to \infty} \log \frac{n+1}{n}$$

$$= \lim_{n \to \infty} \log\left(1 + \frac{1}{n}\right)$$

$$= 0$$

Thus all the three conditions of Leibnitz's test are being satisfied by the given series. Hence by Leibnitz's test, the given series is convergent.

Ex. 15. *Test the series*

$$\log\frac{1}{2} - \log\frac{2}{3} + \log\frac{3}{4} - \log\frac{4}{5} + \ldots$$

Sol. We observe here that

(1) the terms are alternately +ve and –ve

(2) each term is numerically less than its preceding term, as

$$\log\left(\frac{1}{2}\right) > \log\left(\frac{2}{3}\right) > \log\left(\frac{3}{4}\right) > \log\left(\frac{4}{5}\right) > \ldots$$

(3) $u_n = \log\left(\dfrac{n}{n+1}\right)$

$$= \log\left(\frac{1}{1 + \frac{1}{n}}\right)$$

$$\therefore \quad \lim_{n \to \infty} u_n = 0.$$

Thus all the three conditions of Leibnitz's test are being satisfied by the given series. Hence the given series is convergent by Leibnitz's test.

Ex. 16. *Test the following series for convergence*

$$\frac{1}{x} - \frac{1}{x+a} + \frac{1}{x+2a} - \frac{1}{x+3a} + \ldots$$

x and a being positive quantities.

Sol. We observe here that

(1) the terms are alternately +ve and –ve

(2) each term is numerically less than its preceding term, as

$$\frac{1}{x} > \frac{1}{x+a} > \frac{1}{x+2a} > \frac{1}{x+3a} > \ldots$$

($\because$ as x and a are +ve quantities)

(3) $\lim_{n \to \infty} u_n$

$$= \lim_{n \to \infty} \frac{1}{x+(n-1)a}$$

$$= 0$$

Thus all the three conditions of Leibnitz's test are being satisfied by the given series. Hence by Leibnitz's test, the given series is convergent.

Ex. 17. *Examine the following series for convergence*

$$1 - \frac{1}{1+a} + \frac{1}{1+2a} - \frac{1}{1+3a} + \ldots \text{ if } a > 1$$

Sol. We observe here that

(1) the terms of the series are alternately +ve and –ve as $a > 1$

(2) each term is numerically less than its preceding term, as

$$1 > \frac{1}{1+a} > \frac{1}{1+2a} > \frac{1}{1+3a} > \ldots \quad (\because a > 1)$$

(3) $\lim_{n \to \infty} u_n$

$$= \lim_{n \to \infty} \frac{1}{1+(n-1)a}$$

$$= 0$$

Thus all the three conditions of Leibnitz's test are being satisfied by the series. Hence by Leibnitz's test, the given series is convergent.

Ex. 18. *Show that the series*

$$\frac{x}{1+x} - \frac{x^2}{1+x^2} + \frac{x^3}{1+x^3} - \frac{x^4}{1+x^4} + \ldots$$

converges if $0 < x < 1$.

Sol. The given series can be written as

$$\frac{1}{1+\frac{1}{x}} - \frac{1}{1+\frac{1}{x^2}} + \frac{1}{1+\frac{1}{x^3}} - \frac{1}{1+\frac{1}{x^4}} + \ldots$$

Now $0 < x < 1$

So, $\frac{1}{x} > 1.$

Thus each term of the series is numerically less than its preceding term, as

$$\frac{1}{1+\frac{1}{x}} > \frac{1}{1+\frac{1}{x^2}} > \frac{1}{1+\frac{1}{x^3}} + \frac{1}{1+\frac{1}{x^4}} > \ldots$$

$$\left(\because \frac{1}{x} < \frac{1}{x^2} < \frac{1}{x^3} < \frac{1}{x^4} < \ldots\right)$$

Also the terms are alternately +ve and –ve as $\frac{1}{x} > 1$,

and $\underset{n \to \infty}{\text{lt.}} u_n$

$$= \underset{n \to \infty}{\text{lt.}} \frac{1}{1+\frac{1}{x^n}}$$

$$= 0. \left(\text{as } \frac{1}{x} > 1\right)$$

Thus all the three conditions of Leibnitz's test are being satisfied by the given series. Hence the given series is convergent by Leibnitz's test.

Ex. 19. *Prove that the series*

$$\frac{1}{xy}-\frac{1}{(x+1)(y+1)}+\frac{1}{(x+2)(y+2)}-\frac{1}{(x+3)(y+3)}+\ldots$$

where x and y are +ve is convergent.

Sol. We are given that x and y are +ve therefore, we observe in the given series that,

(1) the terms of the series are alternately +ve and –ve,

(2) $\because \quad x + 1 > x$

and $y + 1 > y$

$\therefore \quad (x + 1)(y + 1) > xy$

$$\therefore \quad \frac{1}{(x+1)(y+1)} < \frac{1}{xy}$$

Again, $x + 2 > x + 1$

$y + 2 > y + 1$

$\therefore \quad (x + 2)(y + 2) > (x + 1)(y + 1)$

$$\therefore \quad \frac{1}{(x+2)(y+2)} < \frac{1}{(x+1)(y+1)}$$

and so on.

Hence each term of the series is numerically less than its preceding terms.

(3) $$\lim_{n\to\infty} u_n = \lim_{n\to\infty} \frac{1}{(x+n-1)(y+n-1)} = \frac{1}{\infty} = 0,$$

Thus all the three conditions of Leibnitz's test are being satisfied by the given series. Hence by Leibnitz's test, the given series is convergent.

Ex. 20. *Prove that the series*

$1 - 2x + 3x^2 - 4x^3 + \ldots\ldots$ *where* $x < 1$ *is convergent.*

Sol. In the given series we observe that,

(1) the terms are alternately +ve and –ve

(2) each term is numerically less than its preceding term, as $x < 1$

(3) $\underset{n\to\infty}{\text{lt.}}\ u_n$

$= \underset{n\to\infty}{\text{lt.}}\ nx^{n-1}$

$= \underset{n\to\infty}{\text{lt.}}\ \frac{1}{x}(nx^n)$

$= \frac{1}{x}\ \underset{n\to\infty}{\text{lt.}}\ (nx^n)$

$= \frac{1}{x}\,.\,0\ (\text{as } x<1)$

$= 0$

Thus all the three conditions of Leibnitz's test are being satisfied.

Hence by Leibnitz's test, the given series is convergent.

1.13 Comparison Test for Positive Term Series

Statement. If the ratio of the corresponding terms of two series, in each of which all the terms are positive, be always a fixed finite non-zero quantity, then the two series are convergent or divergent simultaneously.

Or

If Σu_n and Σv_n be two positive term series such that

$$\underset{n\to\infty}{\text{lt.}}\left(\frac{u_n}{v_n}\right)$$

is a fixed finite non-zero quantity, then both the series will converge or diverge simultaneously, i.e., the two series are either both convergent or both divergent.

Proof. Let $\Sigma u_n = u_1 + u_2 + u_3 + \ldots\ldots$

and $\Sigma v_n = v_1 + v_2 + v_3 + \ldots\ldots$ be two given positive term series. We are given that

$\frac{u_1}{v_1}, \frac{u_2}{v_2}, \frac{u_3}{v_3}, \ldots\ldots$ are all finite.

Let l_1 and l_2 be respectively the greatest and least values of these

Let l_1 and l_2 be respectively the greatest and least values of these ratios then,

$$l_1 \geq \frac{u_1}{v_1} \geq l_2 \quad \text{or} \quad v_1 l_1 \geq u_1 \geq v_1 l_2$$

$$l_1 \geq \frac{u_2}{v_2} \geq l_2 \quad \text{or} \quad v_2 l_1 \geq u_2 \geq v_2 l_2$$

$$l_1 \geq \frac{u_3}{v_3} \geq l_2 \quad \text{or} \quad v_3 l_1 \geq u_3 \geq v_3 l_2$$

. . . . . .

. . . . . .

. . . . . .

. . . . . .

and so on.

Adding the above results simultaneously, we get

$$v_1 l_1 + v_2 l_1 + v_3 l_1 + \ldots\ldots \geq u_1 + u_2 + u_3 + \ldots\ldots\ldots$$
$$\geq v_1 l_2 + v_2 l_2 + v_3 l_2 + \ldots\ldots$$

or, $(v_1 + v_2 + v_3 + \ldots)l_1 \geq u_1 + u_2 + u_3 + \ldots\ldots \geq (v_1 + v_2 + v_3 + \ldots)l_2$

or, $l_1 \geq \dfrac{u_1 + u_2 + u_3 + \ldots\ldots}{v_1 + v_2 + v_3 + \ldots\ldots} \geq l_2$

Since all the ratios are finite,

Therefore l_1 and l_2 are also finite. Hence $\dfrac{u_1 + u_2 + u_3 + \ldots}{v_1 + v_2 + v_3 + \ldots}$ which lies between l_1 and l_2 must be finite. Let it be l.

$$\therefore \quad \frac{u_1 + u_2 + u_3 + \ldots}{v_1 + v_2 + v_3 + \ldots} = l$$

or, $$u_1 + u_2 + u_3 + \ldots = l(v_1 + v_2 + v_3 + \ldots)$$

or, $$\Sigma u_n = l \Sigma v_n$$

Thus Σu_n is convergent or divergent,

according as Σv_n is convergent or divergent.

1.14 Working Rule for Comparison Test

First of all we find out the nth term of the given series whose convergence or divergence is to be tested and denote it by u_n. Then for the sake of comparison, we construct a suitable test series whose convergence or divergence is already known to us. This series is called the Auxiliary Series. We find its nth term and denote it by v_n.

Then we determine $\underset{n \to \infty}{\text{lt.}} \frac{u_n}{v_n}$. If this limit comes out to be a fixed finite non-zero quality, then we say that the comparison test can be applied to test the convergence or divergence of the given series.

If it is not so, then we cannot apply the comparison test. If the comparison test is applicable, then, we conclude that the given series will be convergent or divergent according as the auxiliary series Σv_n is convergent or divergent.

1.15 General Method for Finding the Suitable Auxiliary Series

Let Σu_n be the given series. Write down u_n, the nth term of the series. Retain the terms containing highest power of n in the numerator and denominator, separately and simultaneously and then neglect any numerical factors present therein. Thus what we get is n^{th} term of the auxiliary series i.e., v_n. For example :

If we have u_n in the form

$$\frac{A_1 n^p + A_2 n^{p-1} + \ldots\ldots}{B_1 n^q + B_2 n^{q-1} + \ldots\ldots}$$

Then the uth term of the auxiliary series will be

$$\frac{n^p}{n^q}$$

$$\text{i.e. } v_n = \frac{1}{n^{q-p}}$$

1.16 p-series Test or Hyper-Harmonic Test

Statement. The infinite series

$$\Sigma \frac{1}{n^p} \text{ i.e.} \frac{1}{1^p} + \frac{1}{2^p} + \frac{1}{3^p} + \frac{1}{4^p} + \ldots\ldots + \frac{1}{n^p} + \ldots.$$

is convergent if $p > 1$ and divergent if $p \leq 1$.

Proof.

Case I. When $p > 1$.

Let the terms of the given series be grouped such that the first second, third, ..., groups contain 1 term, 2 terms, 2^2 terms,...., respectively. Let S denote the sum of the series. Then, we have

$$S = \frac{1}{1^p} + \left(\frac{1}{2^p} + \frac{1}{3^p}\right) + \left(\frac{1}{4^p} + \frac{1}{5^p} + \frac{1}{6^p} + \frac{1}{7^p}\right)$$

$$+ \left(\frac{1}{8^p} + \frac{1}{9^p} + \ldots + \frac{1}{15^p}\right) + \ldots$$

Now $\because \quad p > 1$

$\therefore \quad 3^p > 2^p$

$\therefore \quad \dfrac{1}{3^p} < \dfrac{1}{2^p}$

$\therefore \quad \dfrac{1}{2^p} + \dfrac{1}{3^p} < \dfrac{1}{2^p} + \dfrac{1}{2^p}$

i.e. $< \dfrac{2}{2^p}$

i.e. $< \dfrac{1}{2^{p-1}}$

Similarly,

$$\frac{1}{4^p} + \frac{1}{5^p} + \frac{1}{6^p} + \frac{1}{7^p}$$

$$< \frac{1}{4^p} + \frac{1}{4^p} + \frac{1}{4^p} + \frac{1}{4^p}$$

i.e. $< \dfrac{4}{4^p}$

i.e. $< \dfrac{1}{4^{p-1}}$

i.e. $< \dfrac{1}{2^{2(p-1)}}$

and $\quad \dfrac{1}{8^p} + \dfrac{1}{9^p} + \ldots + \dfrac{1}{15^p}$

$$< \frac{1}{8^p} + \frac{1}{8^p} + \ldots + \frac{1}{8^p}$$

i.e. $< \dfrac{8}{8^p}$

i.e. $< \dfrac{1}{8^{p-1}}$

i.e. $< \dfrac{1}{2^{3(p-1)}}$

and so on.

Thus, $$S < 1 + \frac{1}{2^{p-1}} + \frac{1}{2^{2(p-1)}} + \frac{1}{2^{3(p-1)}} + \ldots$$

Now the series on right-hand side is a geometric series whose common ratio is $\dfrac{1}{2^{p-1}}$ and as $p > 1$, i.e., $p - 1 > 0$, $\dfrac{1}{2^{p-1}} < 1$ and > 0.

Hence the series on R.H.S. is convergent.

Since S is less than a convergent series,

therefore the given series is convergent when $p > 1$.

Case II. When $p = 1$, then the given series becomes

$$1 + \frac{1}{2} + \frac{1}{3} + \frac{1}{4} + \frac{1}{5} + \ldots$$

Let the terms of the given series be grouped such that first, second, third, fourth, fifth,.... groups contain. 1 term, 1 term, 2 terms, 2^2 terms, 2^3 terms,..... respectively. Then we have

$$S = 1 + \frac{1}{2} + \left(\frac{1}{3} + \frac{1}{4}\right) + \left(\frac{1}{5} + \frac{1}{6} + \frac{1}{7} + \frac{1}{8}\right)$$

$$+ \left(\frac{1}{9} + \frac{1}{10} + \ldots\ldots\ldots \frac{1}{16}\right) + \ldots\ldots.$$

Now, since $3 < 4$

$$\therefore \quad \frac{1}{3} > \frac{1}{4}$$

$$\therefore \quad \frac{1}{3} + \frac{1}{4} > \frac{1}{4} + \frac{1}{4}$$

or, $\frac{1}{3}+\frac{1}{4}>\frac{1}{2}$

Similarly, $\frac{1}{5}+\frac{1}{6}+\frac{1}{7}+\frac{1}{8}$

$$>\frac{1}{8}+\frac{1}{8}+\frac{1}{8}+\frac{1}{8}$$

$$>\frac{4}{8}$$

$$>\frac{1}{2}$$

and so on.

$$\therefore \quad S>1+\frac{1}{2}+\frac{1}{2}+\frac{1}{2}+....$$

The series on R.H.S. is divergent because each term of this series is +ve and greater than a fixed finite quantity $\frac{1}{4}$.

Thus, the sum of the given series S is greater than a divergent series.

Hence the given series is divergent when $p = 1$,

i.e. the series $\Sigma\frac{1}{n}$ is a divergent. The series $\Sigma\frac{1}{n}=1+\frac{1}{2}+\frac{1}{3}+...$ is known as Harmonic series. Thus a harmonic series is divergent.

Case III. When $p < 1$ or negative.

In this case,

$$2^p < 2 \quad \therefore \quad \frac{1}{2^p}>\frac{1}{2}$$

$$3^p < 3 \quad \therefore \quad \frac{1}{3^p}>\frac{1}{3}$$

$$4^p < 4 \quad \therefore \quad \frac{1}{4^p}>\frac{1}{4}$$

and so on.

$$\therefore \quad S = 1 + \frac{1}{2^p} + \frac{1}{3^p} + \frac{1}{4^p} + \ldots$$

$$> 1 + \frac{1}{2} + \frac{1}{3} + \frac{1}{4} + \ldots$$

The series on R.H.S. is divergent by virtue of case II. Hence sum of the given series S is greater than a divergent series.

Hence the given series is divergent.

Note : Two series are very commonly used as auxiliary series,

(1) Geometric series

(2) $\Sigma \frac{1}{n^p}$ or p–series.

Illustrative Examples

Ex. 1. *Test the series*

$$1 + \frac{1}{3} + \frac{1}{5} + \frac{1}{7} + \ldots$$

Sol. Let $\Sigma u_n = 1 + \frac{1}{3} + \frac{1}{5} + \frac{1}{7} + \ldots$

Then, $u_n = \frac{1}{2n-1}$

$v_n = \frac{1}{n}$ where v_n is n^{th} term of the auxiliary series Σv_n.

$$\frac{u_n}{v_n} = \frac{n}{2n-1}$$

$$\therefore \lim_{n \to \infty} \frac{u_n}{v_n} = \lim_{n \to \infty} \frac{n}{2n-1}$$

$$= \lim_{n \to \infty} \frac{1}{2 - \frac{1}{n}}$$

$$= \frac{1}{2} \text{ which is a fixed finite non-zero quantity.}$$

Hence comparison test can be applied.

Now by p-test we know that the series $\Sigma \frac{1}{n}$ is divergent as $p = 1$.

Hence the given series Σu_n is also divergent by comparison test.

Ex. 2. *Test the series*

$$1 + \frac{1}{\sqrt{2}} + \frac{1}{\sqrt{3}} + \frac{1}{\sqrt{4}} + \ldots$$

Sol. Let $\Sigma u_n = 1 + \frac{1}{\sqrt{2}} + \frac{1}{\sqrt{3}} + \frac{1}{\sqrt{4}} + \ldots$

Then, $u_n = \frac{1}{\sqrt{n}}$

$v_n = \frac{1}{\sqrt{n}}$ where v_n is n^{th} term of the auxiliary series $\Sigma\, v_n$,

$$\frac{u_n}{v_n} = 1$$

$$\therefore \underset{n \to \infty}{\text{lt.}} \frac{u_n}{v_n} = 1 \text{ which is a fixed finite non-zero quantity.}$$

Hence comparison test can be applied.

Now by p-test, we know that the series $\Sigma \frac{1}{\sqrt{n}}$ is divergent as $p\left(= \frac{1}{2}\right) < 1$ hence the given series is divergent by comparison test.

Ex. 3. *Test the series*

$$\frac{14}{1^3} + \frac{24}{2^3} + \frac{34}{3^3} + \ldots + \frac{10n+4}{n^3} + \ldots$$

Sol. Let $\Sigma u_n = \frac{14}{1^3} + \frac{24}{2^3} + \frac{34}{3^3} + \ldots + \frac{10n+4}{n^3} + \ldots$

Then, $u_n = \frac{10n+4}{n^3}$

$v_n = \frac{n}{n^3} = \frac{1}{n^2}$ where v_n is the nth term of the auxiliary series Σv_n.

$$\frac{u_n}{v_n} = \frac{(10n+4)}{n^3}.n^2$$

$$= \frac{10n+4}{n}$$

$$= \left(10 + \frac{4}{n}\right)$$

$$\therefore \quad \underset{n \to \infty}{\text{lt.}} \frac{u_n}{v_n} = \underset{n \to \infty}{\text{lt.}} \left(10 + \frac{4}{n}\right)$$

$= 10$ which is a fixed finite non-zero quantity.

Hence comparison test can be applied.

But by p-test we know that the auxiliary series $\Sigma \frac{1}{n^2}$ is convergent as $p(=2) > 1$.

Hence by comparison test, the given series Σu_n is also convergent.

Ex. 4. *Test the series*

$$\frac{1}{\sqrt{2}-1} + \frac{1}{\sqrt{3}-1} + \frac{1}{\sqrt{4}-1} + \ldots$$

Sol. Let $\Sigma u_n = \frac{1}{\sqrt{2}-1} + \frac{1}{\sqrt{3}-1} + \frac{1}{\sqrt{4}-1} + \ldots$

Then, $u_n = \frac{1}{\sqrt{n+1}-1}$

$v_n = \frac{1}{\sqrt{n}}$ where v_n is the nth term of the auxiliary series Σu_n.

$$\frac{u_n}{v_n} = \frac{\sqrt{n}}{\sqrt{n+1}-1}$$

$$= \frac{1}{\sqrt{1+\frac{1}{n}-\frac{1}{\sqrt{n}}}}$$

$\therefore \lim_{n \to \infty} \frac{u_n}{v_n} = 1$ which is a fixed finite non-zero quantity.

Hence comparison test can be applied.

By p-test we know that the auxiliary series $\Sigma \frac{1}{\sqrt{n}}$ is divergent as $p\left(=\frac{1}{2}\right)<1$.

Hence the given series Σu_n is also divergent by comparison test.

Ex. 5. *Test the series*

$$\frac{2}{1}+\frac{3}{8}+\frac{4}{27}+\frac{5}{64}+...$$

Sol. Let $\Sigma u_n = \frac{2}{1}+\frac{3}{8}+\frac{4}{27}+\frac{5}{64}+...$

Then, $u_n = \frac{n+1}{n^3}$

$v_n = \frac{n}{n^3} = \frac{1}{n^2}$ where v_n is n^{th} term of the auxiliary series Σv_n.

$$\frac{u_n}{v_n} = \frac{n+1}{n^3}n^2$$

$$= \frac{n+1}{n}$$

$$= 1 + \frac{1}{n}$$

$$\therefore \quad \lim_{n \to \infty} \frac{u_n}{v_n} = \lim_{n \to \infty} \left(1 + \frac{1}{n}\right)$$

$= 1$ which is a fixed finite non-zero quantity.

Hence comparison test can be applied.

By p-test we know that the auxiliary series $\Sigma v_n = \Sigma \frac{1}{n^2}$ is convergent as $p\ (=2) > 1$ hence by comparison test, the given series is convergent.

Ex. 6. *Test the series*

$$\frac{1}{2} + \frac{\sqrt{2}}{5} + \frac{\sqrt{3}}{10} + \ldots + \frac{\sqrt{n}}{n^2 + 1} + \ldots$$

Sol. Let $\Sigma u_n = \frac{1}{2} + \frac{\sqrt{2}}{5} + \frac{\sqrt{3}}{10} + \ldots + \frac{\sqrt{n}}{n^2 + 1} + \ldots$

Then, $u_n = \frac{\sqrt{n}}{n^2 + 1}$

$$v_n = \frac{\sqrt{n}}{n^2} = \frac{1}{n^{3/2}}$$

where v_n is n^{th} term of the auxiliary series Σv_n.

$$\frac{u_n}{v_n} = \frac{\sqrt{n}}{n^2 + 1} \cdot \frac{n^2}{\sqrt{n}}$$

$$= \frac{1}{1 + \frac{1}{n^2}}$$

$$\therefore \quad \lim_{n \to \infty} \frac{u_n}{v_n} = \lim_{n \to \infty} \left(1 + \frac{1}{n^2}\right)$$

$= 1$ which is a fixed finite non-zero quantity.

Hence comparison test can be applied.

Now by p-test we know that the auxiliary series.

$\Sigma u_n = \Sigma \frac{1}{n^{3/2}}$ is convergent as $p\left(=\frac{3}{2}\right)>1$ hence by comparison test, the given series Σu_n is also convergent.

Ex. 7. *Test the series*

$$\frac{1}{5}+\frac{\sqrt{2}}{7}+\frac{\sqrt{3}}{9}+\frac{\sqrt{4}}{11}+\ldots$$

Sol. Let, $\Sigma u_n = \frac{1}{5}+\frac{\sqrt{2}}{7}+\frac{\sqrt{3}}{9}+\frac{\sqrt{4}}{11}+\ldots$

Then, $u_n = \frac{\sqrt{n}}{2n+3}$

$$\left\{\begin{array}{ll}\textbf{Hint :} & 5, 7, 9, 11, \ldots\ldots \text{ are in A.P.} \\ \text{Its } n^{\text{th}} \text{ term} & = 5+(n-1)^2 \\ & = 5+2n-2 \\ & = 2n+3\end{array}\right\}$$

$$v_n = \frac{\sqrt{n}}{n^{\frac{1}{2}}} = \frac{1}{n^{\frac{1}{2}}}$$

where v_n is n^{th} term of the auxiliary series Σv_n.

$$\frac{u_n}{v_n} = \frac{\sqrt{n}}{2n+3}\sqrt{n} = \frac{n}{2n+3} = \frac{1}{2+\frac{3}{n}}$$

$$\therefore \quad \underset{n\to\infty}{\text{lt.}} \frac{u_n}{v_n} = \underset{n\to\infty}{\text{lt.}} \left(\frac{1}{2+\frac{3}{n}}\right)$$

$$= \frac{1}{2}$$

which is a fixed finite non-zero quantity.

Hence comparison test can be applied.

Now by p-test we know that the auxiliary series $\Sigma \frac{1}{n^{\frac{1}{2}}}$ is divergent as $p\left(=\frac{1}{2}\right)<1$.

Hence the given series is divergent by comparison test.

Ex. 8. *Test the series*

$$\sqrt{\frac{1}{2^3}}+\sqrt{\frac{2}{3^3}}+\sqrt{\frac{3}{4^3}}+\ldots$$

Sol. Let $\quad \Sigma u_n=\sqrt{\frac{1}{2^3}}+\sqrt{\frac{2}{3^3}}+\sqrt{\frac{3}{4^3}}+\ldots$

Then, $\quad u_n=\sqrt{\frac{n}{(n+1)^3}}$

$$v_n=\sqrt{\frac{n}{n^3}}=\sqrt{\frac{1}{n^2}}=\frac{1}{n}$$

where v_n is n^{th} term of the auxiliary series Σv_n.

$$\frac{u_n}{v_n}=\sqrt{\frac{n}{(n+1)^3}}\,.\,n^2$$

$$=\sqrt{\frac{1}{\left(1+\frac{1}{n}\right)^3}}$$

$$\therefore \quad \underset{n\to\infty}{\text{lt.}}\ \frac{u_n}{v_n}=1$$

which is a fixed finite non-zero quantity.

Hence comparison test can be applied.

But by p-test we know that the auxiliary series $\Sigma\frac{1}{n}$ is divergent as $p=1$. Hence the given series is also divergent by comparison test.

Ex. 9. *Test the series*

$$1+\frac{1}{2^2}+\frac{2^2}{3^3}+\frac{3^3}{4^4}$$

Sol. Leave the first term which will not affect the nature of the series and let

$$\Sigma u_n = \frac{1^1}{2^2}+\frac{2^2}{3^3}+\frac{3^3}{4^4}+\ldots$$

Then, $$u_n = \frac{n^n}{(n+1)^{n+1}}$$

$$v_n = \frac{n^n}{n^{n+1}} = \frac{1}{n}$$

where v_n is n^{th} term of the auxiliary series Σv_n.

$$\frac{u_n}{v_n} = \frac{n^{n+1}}{(n+1)^{n+1}}$$

$$= \left(\frac{n}{n+1}\right)^{n+1}$$

$$= \frac{1}{\left(1+\frac{1}{n}\right)^{n+1}}$$

$$= \frac{1}{\left(1+\frac{1}{n}\right)^{n}}\cdot\frac{1}{1+\frac{1}{n}}$$

$$\therefore \quad \underset{n\to\infty}{\text{lt.}} \frac{u_n}{v_n} = \frac{1}{e}$$

which is a fixed finite non-zero quantity.

Hence comparison test can be applied.

But by p-test we know that the auxiliary series $\Sigma\frac{1}{n}$ is divergent as $p = 1$.

Hence the given series is also divergent by comparison test.

Ex. 10. *Test the series*

$$\frac{\sqrt{1}}{1+\sqrt{1}}+\frac{\sqrt{2}}{2+\sqrt{2}}+\frac{\sqrt{3}}{3+\sqrt{3}}+\ldots$$

Sol. Let $\Sigma u_n = \frac{\sqrt{1}}{1+\sqrt{1}}+\frac{\sqrt{2}}{2+\sqrt{2}}+\frac{\sqrt{3}}{3+\sqrt{3}}+\ldots$

Then, $u_n = \frac{\sqrt{n}}{n+\sqrt{n}}$

$v_n = \frac{\sqrt{n}}{n} = \frac{1}{n^{\frac{1}{2}}}$

where v_n is n^{th} term of the auxiliary series Σv_n.

$$\frac{u_n}{v_n} = \frac{n}{n+\sqrt{n}}$$

$$\frac{1}{1+\frac{1}{n^{\frac{1}{2}}}}$$

$\therefore \lim_{n\to\infty} \frac{u_n}{v_n} = 1$

which is a fixed finite non-zero quantity.

Hence comparison test can be applied.

But by p-test we know that the series $\Sigma\frac{1}{n^{\frac{1}{2}}}$ is divergent as $p\left(=\frac{1}{2}\right)<1$.

Hence by comparison test, the given series is divergent.

Ex. 11. *Test the series*

$$\frac{1}{1^p}+\frac{1}{3^p}+\frac{1}{5^p}+\frac{1}{7^p}+\ldots$$

Sol. Let $\Sigma u_n = \frac{1}{1^p}+\frac{1}{3^p}+\frac{1}{5^p}+\frac{1}{7^p}+\ldots$

Then, $u_n = \dfrac{1}{(2n-1)^p}$

$v_n = \dfrac{1}{n^p}$ when v_n is n^{th} term of the auxiliary series Σv_n,

$$\frac{u_n}{v_n} = \frac{n^p}{(2n-1)^p} = \frac{1}{\left(2-\frac{1}{n}\right)^p}$$

$\therefore \lim_{n \to \infty} \dfrac{u_n}{v_n} = \dfrac{1}{2^p}$ which is a fixed finite non-zero quantity.

Hence comparison test can be applied.

But by p-test we know that the auxiliary series, $\Sigma \dfrac{1}{n^p}$ is convergent if $p > 1$ and divergent if $p \le 1$. Hence the given series will also be convergent if $p > 1$ and divergent if $p \le 1$.

Ex. 12. *Test the series*

$$\frac{2}{1^p} + \frac{3}{2^p} + \frac{4}{3^p} + \frac{5}{4^p} + \ldots$$

Sol. Let $\Sigma u_n = \dfrac{2}{1^p} + \dfrac{3}{2^p} + \dfrac{4}{3^p} + \dfrac{5}{4^p} + \ldots$

Then, $u_n = \dfrac{n+1}{n^p}$

$v_n = \dfrac{n}{n^p} = \dfrac{1}{n^{p-1}}$ where v_n is the nth term of the auxiliary series Σv_n.

$$\frac{u_n}{v_n} = \frac{n+1}{n^p} n^{p-1}$$

$$= \frac{n+1}{n} = 1 + \frac{1}{n}$$

$\therefore \lim_{n \to \infty} \dfrac{u_n}{v_n} = 1$ which is a fixed finite non-zero quantity.

Hence comparison test can be applied.

But by p-test we know that the auxiliary series $\Sigma \frac{1}{n^{p-1}}$ will be convergent if $p-1>1$ i.e. $p>2$ and divergent if $p-1 \le 1$ i.e. $p \le 2$. Hence the given series Σu_n will also be convergent when $p>2$ and divergent when $p \le 2$.

Ex. 13. *Test the series*

$$\sum_{n=1}^{\infty} \frac{1}{(3n)^p};\ p>1$$

Sol. Let $\Sigma u_n = \sum_{n=1}^{\infty} \frac{1}{(3n)^p}$

Then, $u_n = \frac{1}{(3n)^p}$

$v_n = \frac{1}{n^p}$ where v_n is the nth term of the auxiliary series Σv_n.

$\frac{u_n}{v_n} = \frac{n^p}{(3n)^p} = \frac{1}{3^p}$ which is a fixed finite non-zero quantity,

Hence comparison test can be applied.

But by p-test we know that $\Sigma \frac{1}{n^p}$ is convergent as $p>1$. Hence the given series is also convergent by comparison test.

Ex. 14. *Test the series*

$$\frac{2^p}{1^q} + \frac{3^p}{2^q} + \frac{4^p}{3^q} + \ldots$$

Sol. Let $\Sigma u_n = \frac{2^p}{1^q} + \frac{3^p}{2^q} + \frac{4^p}{3^q} + \ldots$

Then, $u_n = \frac{(n+1)^p}{n^q}$

$$v_n = \frac{n^p}{n^q}$$

where v_n is the nth term of the auxiliary series Σv_n.

$$\frac{u_n}{v_n} = \frac{(n+1)^p}{n^q} \cdot n^{q-p}$$

$$= \left(\frac{n+1}{n}\right)^p = \left(1+\frac{1}{n}\right)^p$$

$\therefore \lim_{n \to \infty} \frac{u_n}{v_n} = 1$ which is a fixed finite non-zero quantity.

Hence comparison test can be applied.

But by p-test we know that the auxiliary series $\Sigma \frac{1}{n^{q-p}}$ is convergent if $q - p > 1$ i.e. $q > p + 1$ and divergent if $q - p \leq 1$ i.e. $q \leq p + 1$.

Hence the given series will also be convergent if $q > p + 1$ and divergent if $q \leq p + 1$.

Ex. 15. *Test the series*

$$\frac{1}{a.1^2 + b} + \frac{2}{a.2^2 + b} + ... + \frac{n}{a.n^2 + b} + ...$$

Sol. Let $\Sigma u_n = \frac{1}{a.1^2 + b} + \frac{2}{a.2^2 + b} + ... + \frac{n}{a.n^2 + b} + ...$

Then, $u_n = \frac{n}{an^2 + b}$

$v_n = \frac{n}{n^2} = \frac{1}{n}$ where v_n is n^{th} term of the auxiliary series Σv_n.

$$\frac{u_n}{v_n} = \frac{n^2}{an^2 + b}$$

$$= \frac{1}{a + \frac{b}{n^2}}$$

$\therefore \lim_{n \to \infty} \frac{u_n}{v_n} = \frac{1}{a}$ which is a fixed finite non-zero quantity.

Hence comparison test can be applied.

Now by *p*-test we know that the auxiliary series $\Sigma \frac{1}{n}$ is divergent as $p = 1$. Hence the given series is also divergent by comparison test.

Ex. 16. *Test the series*

$$\frac{1}{a^2+1} + \frac{1}{2a^2+1} + \frac{1}{3a^2+1} + \ldots$$

Sol. Let $\Sigma u_n = \frac{1}{a^2+1} + \frac{1}{2a^2+1} + \frac{1}{3a^2+1} + \ldots$

Then, $u_n = \frac{1}{na^2+1}$

$v_n = \frac{1}{n}$ where v_n is n^{th} term of the auxiliary series Σv_n.

$$\frac{u_n}{v_n} = \frac{1}{na^2+1} = \frac{1}{a^2 + \frac{1}{n}}$$

$\therefore \lim_{n \to \infty} \frac{u_n}{v_n} = \frac{1}{a^2}$ which is a fixed finite non-zero quantity.

Hence comparison test can be applied.

But by *p*–test know that the auxiliary series $\Sigma \frac{1}{n}$ is divergent as $p = 1$. Hence the given series is also divergent by comparison test.

Ex. 17. *Test the series*

$$\frac{1}{a+b} + \frac{1}{a+2b} + \frac{1}{a+3b} + \ldots + \frac{1}{a+nb} + \ldots$$

Sol. Let $\Sigma u_n = \frac{1}{a+b} + \frac{1}{a+2b} + \frac{1}{a+3b} + \ldots + \frac{1}{a+nb} + \ldots$

Then, $u_n = \frac{1}{a+nb}$

$$v_n = \frac{1}{n} \text{ what } v_n \text{ is } n^{\text{th}} \text{ term of the auxiliary series } \Sigma v_n.$$

$$\frac{u_n}{v_n} = \frac{n}{a+nb} = \frac{1}{b+\frac{a}{n}}$$

$$\therefore \quad \underset{n\to\infty}{\text{lt.}} \frac{u_n}{v_n} = \frac{1}{b} \text{ which is a fixed finite non-zero quantity.}$$

Hence comparison test can be applied,

But by p-test we know that the auxiliary series $\Sigma \frac{1}{n}$ is divergent as $p = 1$. Hence the given series is also divergent by comparison test.

Ex. 18. *Test the series*

$$\frac{1}{a(a+b)} + \frac{1}{(a+2b)(a+3b)} + \frac{1}{(a+4b)(a+5b)} + \ldots$$

Sol. Let $\Sigma u_n = \frac{1}{a(a+b)} + \frac{1}{(a+2b)(a+3b)} + \frac{1}{(a+4b)(a+5b)} + \ldots.$

Then, $u_n = \frac{1}{\{a+(n-1)2b\}\{a+(2n-1)b\}}$

or, $u_n = \frac{1}{(a-2b+2nb)(a-b+2nb)}$

$$v_n = \frac{1}{n^2} \text{ where } v_n \text{ is } n^{\text{th}} \text{ term of the auxiliary series } \Sigma v_n.$$

$$\frac{u_n}{v_n} = \frac{n^2}{(a-2b+2nb)(a-b+2nb)}$$

$$= \frac{1}{\left(2b+\frac{a-2b}{n}\right)\left(2b+\frac{a-b}{n}\right)}$$

$$\underset{n\to\infty}{\text{lt.}} \frac{u_n}{v_n} = \frac{1}{4b^2} \text{ which is a fixed finite non-zero quantity.}$$

Hence comparison test can be applied.

But by p-test we know that the auxiliary series $\Sigma v_n = \Sigma \frac{1}{n^2}$ is convergent as $p\ (= 2) > 1$. Hence the given series is also convergent by comparison test.

Ex. 19. *Test the series*

$$1 + \frac{1+2}{1+2^2} + \frac{1+3}{1+3^2} + \ldots$$

Sol. The given series can be written as

$$\frac{1+1}{1+1^2} + \frac{1+2}{1+2^2} + \frac{1+3}{1+3^2} + \ldots$$

Let the series by denoted by Σu_n.

Then, $u_n = \dfrac{1+n}{1+n^2}$

$\therefore \quad v_n = \dfrac{n}{n^2} = \dfrac{1}{n}$ where v_n is n^{th} term of the auxiliary series Σv_n.

$$\frac{u_n}{v_n} = \frac{n(1+n)}{1+n^2}$$

$$= \frac{n^2+n}{n^2+1}$$

$$= \frac{1+\frac{1}{n}}{1+\frac{1}{n^2}}$$

$\therefore \quad \underset{n \to \infty}{\text{lt.}} \dfrac{u_n}{v_n} = 1$ which is a fixed finite non-zero quantity.

Hence comparison test can be applied.

But by p-test we know that the auxiliary series $\Sigma \frac{1}{n}$ is divergent as $p = 1$. Hence by comparison test the given series Σu_n is also divergent.

Ex. 20. *Test the series*

$$\frac{1}{1\cdot 2\cdot 3}+\frac{1}{2\cdot 3\cdot 4}+\frac{5}{3\cdot 4\cdot 5}+\ldots$$

Sol. Let the given series be denoted by Σu_n

Then, $u_n = \dfrac{2n-1}{n(n+1)(n+2)}$

$\therefore \quad v_n = \dfrac{n}{n^3} = \dfrac{1}{n^2}$ where v_n is the nth term of the auxiliary series.

$$\frac{u_n}{v_n} = \frac{(2n-1)n^2}{n(n+1)(n+2)}$$

$$= \frac{(2n-1)n}{(n+1)(n+2)}$$

$$= \frac{\left(2-\frac{1}{n}\right)}{\left(1+\frac{1}{n}\right)\left(1+\frac{2}{n}\right)}$$

$\therefore \quad \lim\limits_{n\to\infty} \dfrac{u_n}{v_n} = 2$ which is a fixed finite non-zero quantity.

Hence comparison test can be applied.

But by p-test we know that the auxiliary series $\Sigma\dfrac{1}{n^2}$ is convergent as p (= 2) >1. Hence by comparison test, the given series is also convergent.

Ex. 21. *Test the series*

$$\frac{\sqrt{3}}{1\cdot 2}+\frac{\sqrt{5}}{3\cdot 4}+\frac{\sqrt{7}}{5\cdot 6}+\frac{\sqrt{9}}{7\cdot 8}+\ldots$$

Sol. Let the given series be denoted by Σu_n. Then

$$v_n = \frac{\sqrt{2n+1}}{(2n-1)(2n)}$$

$$\therefore \quad v_n = \frac{\sqrt{n}}{n^2} = \frac{1}{n^{3/2}}$$

where v_n is n^{th} term of the auxiliary series Σv_n.

$$\therefore \quad \frac{u_n}{v_n} = \frac{\sqrt{2n+1}}{(2n-1)2n} \cdot n^{3/2}$$

$$= \frac{\sqrt{2n+1}\sqrt{n}}{(2n-1)2}$$

$$= \frac{\sqrt{2+\frac{1}{n}}}{\left(2-\frac{1}{n}\right) \cdot 2}$$

$$\therefore \quad \underset{n \to \infty}{\text{lt.}} \frac{u_n}{v_n} = \frac{1}{2}$$

which is a fixed finite non-zero quantity.

Hence comparison test can be applied.

But by p-test, we know that the auxiliary series $\Sigma \frac{1}{n^{3/2}}$ is convergent as p (= 3/2) >1. Hence by comparison test, the given series Σu_n is also convergent.

Ex. 22. *Test the series*

$$\frac{1}{(x-d)^2} + \frac{1}{(x-2d)^2} + \frac{1}{(x-3d)^2} + \ldots$$

Sol. Let the given series be denoted by Σu_n then

$$u_n = \frac{1}{(x-nd)^2}$$

$v_n = \frac{1}{n^2}$ where v_n is n^{th} term of the auxiliary series Σv_n.

$$\therefore \quad \frac{u_n}{v_n} = \frac{n^2}{(x-nd)^2}$$

$$= \frac{1}{\left(\frac{x}{n} - d\right)^2}$$

$$\therefore \quad \underset{n \to \infty}{\text{lt.}} \frac{u_n}{v_n} = \frac{1}{d^2}$$

which is a fixed finite non-zero quantity.

Hence comparison test can be applied.

But by p-test, we know that the auxiliary series $\Sigma \frac{1}{n^2}$ is convergent as $p(=2) > 1$ and hence by comparison test, the given series Σu_n is also convergent.

Test for convergence or divergence the series whose general terms (n^{th} terms) are given below :

Examples :

23. $\left(\frac{n^2 + 2}{2n^3 + 15}\right)^{1/3}$

24. $\frac{(2n^2 - 1)^{1/3}}{(3n^3 + 2n + 5)^{1/4}}$

25. $\frac{n^p}{(n+1)^{p+q}}$

26. $\sqrt{n+1} - \sqrt{n}$

27. $\frac{\sqrt{n+1} - \sqrt{n}}{n^p}$

28. $\sqrt{n^2 + 1} - n$

29. $\sqrt{n+1} - \sqrt{n-1}$

30. $\sqrt{n^2 + 1} - \sqrt{n^2 - 1}$

31. $\sqrt{n^3 + 1} - \sqrt{n^3}$

32. $\dfrac{\sqrt{n^2+1}-n}{n^p}$

33. $\sqrt{n^3+1}-\sqrt{n^3-1}$

34. $\sqrt{n^4+1}-n^2$

35. $\sqrt{n^4+1}-\sqrt{n^4-1}$

36. $\sqrt[3]{n^3+1}-n$

37. $\dfrac{\sqrt{n^2+n+1}-\sqrt{n^2-n+1}}{n}$

Solutions

Sol. 23. Let the given series be denoted by Σu_n. Then

$$u_n = \left(\frac{n^2+2}{2n^3+15}\right)^{1/3}$$

$$\therefore \qquad v_n = \left(\frac{n^2}{n^3}\right)^{1/3} = \frac{1}{n^{1/3}}$$

where v_n is n^{th} term of the auxiliary series Σv_n.

$$\frac{u_n}{v_n} = \left(\frac{n^3+2n}{2n^3+15}\right)^{1/3}$$

$$= \left[\frac{1+\dfrac{2}{n^2}}{2+\dfrac{15}{n^3}}\right]^{1/3}$$

$$\therefore \quad \underset{n\to\infty}{\text{lt.}} \frac{u_n}{v_n} = \frac{1}{2^{1/3}}$$

which is a fixed finite non-zero quantity.

Hence comparison test can be applied.

But $\Sigma u_n = \Sigma \frac{1}{n^{1/3}}$ is divergent as $p\left(=\frac{1}{3}\right) < 1$ by p-test.

Hence Σu_n is also divergent by comparison test.

Sol. 24. Let the given series be denoted by Σu_n. Then

$$u_n = \frac{\left(2n^2-1\right)^{1/3}}{\left(3n^2+2n+5\right)^{1/4}}$$

$$\therefore \qquad v_n = \frac{\left(n^2\right)^{1/3}}{\left(n^3\right)^{1/4}} = \frac{n^{2/3}}{n^{3/4}} = \frac{1}{n^{1/12}}$$

where v_n is the nth term of the auxiliary series Σv_n,

$$\frac{u_n}{v_n} = \frac{\left(2n^2-1\right)^{1/3}.n^{1/12}}{\left(3n^3+2n+5\right)^{1/4}}$$

$$= \frac{\left(2n^2-1\right)^{1/3}}{\left(3n^3+2n+5\right)^{1/4}}.\frac{\left(n^3\right)^{1/4}}{\left(n^2\right)^{1/3}}$$

$$= \frac{\left(2-\frac{1}{n^2}\right)^{1/3}}{\left(3+\frac{2}{n^2}+\frac{5}{n^3}\right)^{1/4}}$$

$$\therefore \qquad \underset{n\to\infty}{\text{lt.}} \frac{u_n}{v_n} = \frac{(2)^{1/3}}{(3)^{1/4}}$$

which is a fixed finite non-zero quantity.

Hence comparison test can be applied.

But $\Sigma v_n = \Sigma \frac{1}{n^{1/12}}$ is divergent as $p\left(=\frac{1}{12}\right) < 1$ by p-test.

Hence Σu_n is also divergent by comparison test.

Sol. 25. Let the given series be denoted by Σu_n. Then

$$u_n = \frac{n^p}{(n+1)^{p+q}}$$

$$\therefore \qquad v_n = \frac{n^p}{n^{p+q}} = \frac{1}{n^q}$$

where v_n is n^{th} term of the auxiliary series Σv_n.

$$\frac{u_n}{v_n} = \frac{n^{p+q}}{(n+1)^{p+q}} = \frac{1}{\left(1+\frac{1}{n}\right)^{p+q}}$$

$$\therefore \qquad \underset{n\to\infty}{\text{lt}} \frac{u_n}{v_n} = 1$$

which is a fixed finite non-zero quantity.

Hence comparison test can be applied.

But $\sum v_n = \Sigma \frac{1}{n^q}$ is convergent when $q > 1$ and divergent when $q \le 1$ by p-test. Hence $\sum u_n$ is also convergent when $q > 1$ and divergent when $q \le 1$ by comparison test.

Sol. 26. Let the given series be denoted by Σu_n. Then

$$u_n = \sqrt{n+1} - \sqrt{n}$$

$$= \sqrt{n\left(1+\frac{1}{n}\right)} - \sqrt{n}$$

$$= \sqrt{n}\left\{\left(1+\frac{1}{n}\right)^{\frac{1}{2}} - 1\right\}$$

$$= \sqrt{n}\left\{1 + \frac{1}{2}.\frac{1}{n} + \frac{\frac{1}{2}\left(\frac{1}{2}-1\right)}{2!}\left(\frac{1}{n}\right)^2 + \ldots\ldots - 1\right\}$$

$$= \sqrt{n}\left\{\frac{1}{2n}-\frac{1}{8n^2}+\ldots\ldots\right\}$$

$$= \frac{1}{2n^{1/2}}-\frac{1}{8n^{3/2}}+\ldots\ldots$$

$$\therefore \quad v_n = \frac{1}{n^{1/2}}$$

where v_n is n^{th} term of the auxiliary series Σv_n.

$$\frac{u_n}{v_n} = \frac{1}{2}-\frac{1}{8n}+\ldots$$

$$\therefore \quad \lim_{n\to\infty} \frac{u_n}{v_n} = \frac{1}{2}$$

which is a fixed finite non-zero quantity.

Hence comparison test can be applied.

But $\Sigma \frac{1}{n^{1/2}}$ is divergent as $p\left(=\frac{1}{2}\right)<1$ by p-test.

Hence Σu_n is also divergent by comparison test.

Sol. 27. Let the given series be denoted by Σu_n. Then,

$$u_n = \frac{\sqrt{n+1}-\sqrt{n}}{n^p}$$

or
$$u_n = \frac{\frac{1}{2n^{1/2}}-\frac{1}{8n^{3/2}}+\ldots}{n^p}$$

(by Ex. 26)

$$= \frac{1}{2n^{p+1/2}}-\frac{1}{8n^{p+3/2}}+\ldots$$

$$\therefore \quad v_n = \frac{1}{n^{p+1/2}}$$

where v_n is n^{th} term of the auxiliary series Σv_n.

$$\frac{u_n}{v_n} = \frac{1}{2} - \frac{1}{8n} + \ldots$$

$$\therefore \lim_{n \to \infty} \frac{u_n}{v_n} = \frac{1}{2}$$

which is a fixed finite non-zero quantity.
Hence comparison test can be applied.

But $\Sigma \frac{1}{n^{p+1/2}}$ is convergent if $p + \frac{1}{2} > 1$ i.e. $p > \frac{1}{2}$ and divergent if $p + \frac{1}{2} \leq 1$ i.e. $p \leq \frac{1}{2}$ by p-test.

Hence Σu_n is also convergent when $p > \frac{1}{2}$ and divergent when $p \leq \frac{1}{2}$ by comparison test.

Sol. 28. Let the given series be denoted by Σu_n. Then,

$$u_n = \sqrt{n^2 + 1} - n$$

$$= \sqrt{n^2\left(1 + \frac{1}{n^2}\right)} - n$$

$$= n\left[\left(1 + \frac{1}{n^2}\right)^{1/2} - 1\right]$$

$$= n\left[1 + \frac{1}{2} \cdot \frac{1}{n^2} + \frac{\frac{1}{2}\left(\frac{1}{2} - 1\right)}{2!} \frac{1}{n^4} + \ldots - 1\right]$$

$$= n\left[\frac{1}{2n^2} - \frac{1}{8n^4} + \ldots\right]$$

$$= \frac{1}{2n} - \frac{1}{8n^3} + \ldots$$

where v_n is n^{th} term of the auxiliary series Σv_n.

$$\therefore \frac{u_n}{v_n} = \frac{1}{2} - \frac{1}{8n^2} + \ldots$$

$$\therefore \quad \underset{n \to \infty}{\text{lt.}} \frac{u_n}{v_n} = \frac{1}{2}$$

which is a fixed finite non-zero quantity.

Hence comparison test can be applied.

But $\Sigma \frac{1}{n}$ is divergent as $p = 1$ by p-test. Hence Σu_n is also divergent by comparison test.

Sol. 29. Let the given series be denoted by Σu_n. Then

$$u_n = \sqrt{n+1} - \sqrt{n-1}$$

$$= \sqrt{n}\left\{\left(1+\frac{1}{n}\right)^{1/2} - \left(1-\frac{1}{n}\right)^{1/2}\right\}$$

$$= \sqrt{n}\left[\left\{1+\frac{1}{2}\cdot\frac{1}{n}+\frac{\frac{1}{2}\left(\frac{1}{2}-1\right)}{2!}\cdot\frac{1}{n^2}+\frac{\frac{1}{2}\left(\frac{1}{2}-1\right)\left(\frac{1}{2}-2\right)}{3!}\cdot\frac{1}{n^3}+\ldots\right\}\right.$$

$$\left.-\left\{1-\frac{1}{2}\cdot\frac{1}{n}+\frac{\frac{1}{2}\left(\frac{1}{2}-1\right)}{2!}\cdot\frac{1}{n^2}-\frac{\frac{1}{2}\left(\frac{1}{2}-1\right)\left(\frac{1}{2}-2\right)}{3!}\cdot\frac{1}{n^3}+\ldots\right\}\right]$$

$$= \sqrt{n}\left[\frac{1}{n}+\frac{1}{8n^3}+\ldots\right]$$

$$= \frac{1}{n^{1/2}}+\frac{1}{8n^{5/2}}+\ldots$$

$$v_n = \frac{1}{n^{1/2}}$$

where v_n is the nth term of the auxiliary series Σv_n.

$$\frac{u_n}{v_n} = 1+\frac{1}{8n^2}+\ldots$$

$$\therefore \quad \underset{n \to \infty}{\text{lt.}} \frac{u_n}{v_n} = 1.$$

which is a fixed finite non-zero quantity.

Hence comparison test can be applied.

But $\Sigma v_n = \Sigma \frac{1}{n^{1/2}}$ is divergent as $p\left(=\frac{1}{2}\right) < 1$ by p-test.

Hence Σu_n is also divergent by comparison test.

Sol. 30. Let the given series be denoted by Σu_n. Then

$$u_n = \sqrt{n^2+1} - \sqrt{n^2-1}$$

or $$u_n = \sqrt{n^2\left(1+\frac{1}{n^2}\right)} - \sqrt{n^2\left(1-\frac{1}{n^2}\right)}$$

$$= n\left[\left(1+\frac{1}{n^2}\right)^{1/2} - \left(1-\frac{1}{n^2}\right)^{1/2}\right]$$

$$= n\left[\left\{1+\frac{1}{2}.\frac{1}{n^2}+\frac{\frac{1}{2}\left(\frac{1}{2}-1\right)}{2!}\frac{1}{n^4}+\frac{\frac{1}{2}\left(\frac{1}{2}-1\right)\left(\frac{1}{2}-2\right)}{3!}\frac{1}{n^6}+\ldots\right\}\right.$$

$$\left.-\left\{1-\frac{1}{2}\frac{1}{n^2}+\frac{\frac{1}{2}\left(\frac{1}{2}-1\right)}{2!}\frac{1}{n^4}-\frac{\frac{1}{2}\left(\frac{1}{2}-1\right)\left(\frac{1}{2}-2\right)}{3!}\frac{1}{n^6}+\ldots\right\}\right]$$

$$= n\left[\frac{1}{n^2}+\frac{1}{8n^6}+\ldots\right]$$

$$= \frac{1}{n}+\frac{1}{8n^5}+\ldots$$

$$v_n = \frac{1}{n}$$

where v_n is n^{th} term of the auxiliary series Σv_n.

$$\frac{u_n}{v_n} = 1+\frac{1}{8n^4}+\ldots$$

$$\therefore \lim_{n\to\infty} \frac{u_n}{v_n} = 1$$

which is a fixed finite non-zero quantity.

Hence comparison test can be applied.

But $\Sigma v_n = \Sigma \frac{1}{n}$ is divergent as $p = 1$ by p-test

Hence Σu_n is also divergent by comparison test.

Sol. 31. Let the given series be denoted by Σu_n. Then

$$u_n = \sqrt{n^3+1} - \sqrt{n^3}$$

$$= \sqrt{n^3\left(1+\frac{1}{n^3}\right)} - \sqrt{n^3}$$

$$= n^{3/2}\left[\left(1+\frac{1}{n^3}\right)^{1/2} - 1\right]$$

$$= n^{3/2}\left[1+\frac{1}{2}\frac{1}{n^3}+\frac{\frac{1}{2}\left(\frac{1}{2}-1\right)}{2!}\frac{1}{n^6}+\ldots-1\right]$$

$$= \frac{1}{2n^{3/2}} - \frac{1}{8}\frac{1}{n^{9/2}} + \ldots$$

$$v_n = \frac{1}{n^{3/2}}$$

where v_n is n^{th} term of the auxiliary series Σv_n.

$$\therefore \lim_{n\to\infty} \frac{u_n}{v_n} = \frac{1}{2}$$

which is a fixed finite non-zero quantity.

Hence comparison test can be applied.

But $\Sigma \frac{1}{n^{3/2}}$ is convergent as $p\ (= 3/2) > 1$ by p-test

Hence Σu_n is also convergent by comparison test.

Sol. 32. Let the given series be denoted by Σu_n. Then

$$u_n = \frac{\sqrt{n^2+1}-n}{n^p}$$

$$= \frac{n\left[\left(1+\frac{1}{n^2}\right)^{1/2}-1\right]}{n^p}$$

$$= \frac{1+\frac{1}{2}\cdot\frac{1}{n^2}+\frac{\frac{1}{2}\left(\frac{1}{2}-1\right)}{2!}\frac{1}{n^4}+\ldots-1}{n^{p-1}}$$

$$= \frac{\frac{1}{2n^2}-\frac{1}{8n^4}+\ldots}{n^{p-1}}$$

$$= \frac{1}{2n^{p+1}}-\frac{1}{8n^{p+3}}+\ldots$$

$$\therefore \quad v_n = \frac{1}{n^{p+1}}$$

where v_n is n^{th} term of the auxiliary series Σv_n.

$$\frac{u_n}{v_n} = \frac{1}{2}-\frac{1}{8n^2}+\ldots$$

$$\therefore \quad \lim_{n\to\infty} \frac{u_n}{v_n} = \frac{1}{2}$$

which is a fixed finite non-zero quantity.

Hence comparison test can be applied.

But $\Sigma \frac{1}{n^{p+1}}$ is convergent if $p + 1 > 1$ i.e. $p > 0$ and divergent if $p+1 \le 1$ i.e. $p \le 0$ by p-test.

Hence Σu_n is also convergent when $p > 0$ and divergent when $p \le 0$ by comparison test.

Sol. 33. Let the given series be denoted by Σu_n. Then

$$u_n = \sqrt{n^3+1} - \sqrt{n^3-1}$$

$$u_n = n^{3/2}\left[\left(1+\frac{1}{n^3}\right)^{1/2} - \left(1-\frac{1}{n^3}\right)^{1/2}\right]$$

$$= n^{3/2}\left[\left\{1+\frac{1}{2}\frac{1}{n^3}+\frac{\frac{1}{2}\left(\frac{1}{2}-1\right)}{2!}\frac{1}{n^6}+\frac{\frac{1}{2}\left(\frac{1}{2}-1\right)\left(\frac{1}{2}-2\right)}{3!}\frac{1}{n^9}+\ldots\right\}\right.$$

$$\left.-\left\{1-\frac{1}{2}\frac{1}{n^3}+\frac{\frac{1}{2}\left(\frac{1}{2}-1\right)}{2!}\frac{1}{n^6}-\frac{\frac{1}{2}\left(\frac{1}{2}-1\right)\left(\frac{1}{2}-2\right)}{3!}\frac{1}{n^9}+\ldots\right\}\right]$$

$$= n^{3/2}\left[\frac{1}{n^3}+\frac{1}{8n^9}+\ldots\right]$$

$$= \frac{1}{n^{3/2}}+\frac{1}{8n^{15/2}}+\ldots$$

$$\therefore\ v_n = \frac{1}{n^{3/2}}$$

where v_n is n^{th} term of the auxiliary series Σv_n.

$$\frac{u_n}{v_n} = 1+\frac{1}{8n^6}+\ldots$$

$$\therefore\ \lim_{n\to\infty}\frac{u_n}{v_n} = 1$$

which is a fixed finite non-zero quantity.

Hence comparison test can be applied.

But $\Sigma\frac{1}{n^{3/2}}$ is convergent as $p = (3/2) > 1$ by p-test. Hence Σu_n is also convergent by comparison test.

Sol. 34. Let the given series be devided by Σu_n. Then

$$u_n = \sqrt{n^4+1} - n^2$$

$$= n^2\left[\left(1+\frac{1}{n^4}\right)^{1/2} - 1\right]$$

$$= n^2\left[1 + \frac{1}{2}\frac{1}{n^4} + \frac{\frac{1}{2}\left(\frac{1}{2}-1\right)}{2!}\frac{1}{n^8} + \ldots - 1\right]$$

$$= \frac{1}{2n^2} - \frac{1}{8n^6} + \ldots$$

$\therefore \quad v_n = \dfrac{1}{n^2}$ where v_n is n^{th} term of the auxiliary series Σv_n.

$$\therefore \quad \frac{u_n}{v_n} = \frac{1}{2} - \frac{1}{8n^4} + \ldots$$

$\therefore \quad \lim\limits_{n\to\infty} \dfrac{u_n}{v_n} = \dfrac{1}{2}$ which is a fixed non-zero finite quantity. Hence comparison test can be applied.

But $\Sigma \dfrac{1}{n^2}$ is convergent as $p\,(=2) > 1$ by p-test.

Hence Σu_n is also convergent by comparison test.

Sol. 35. Let the given series be denoted by Σu_n. Then

$$u_n = \sqrt{n^4+1} - \sqrt{n^4-1}$$

$$= n^2\left[\left(1+\frac{1}{n^4}\right)^{1/2} - \left(1-\frac{1}{n^4}\right)^{1/2}\right]$$

$$= n^2\left[\left\{1 + \frac{1}{2}\frac{1}{n^4} + \frac{\frac{1}{2}\left(\frac{1}{2}-1\right)}{2!}\frac{1}{n^8} + \frac{\frac{1}{2}\left(\frac{1}{2}-1\right)\left(\frac{1}{2}-2\right)}{3!}\frac{1}{n^{12}} + \ldots\right\}\right.$$

$$-\left\{1-\frac{1}{2}\frac{1}{n^4}+\frac{\frac{1}{2}\left(\frac{1}{2}-1\right)}{2!}\frac{1}{n^8}-\frac{\frac{1}{2}\left(\frac{1}{2}-1\right)\left(\frac{1}{2}-2\right)}{3!}\frac{1}{n^{12}}+\ldots\right\}\Bigg]$$

$$=n^2\left[\frac{1}{n^4}+\frac{1}{8n^{12}}+\ldots\right]$$

$$=\frac{1}{n^2}+\frac{1}{8n^{10}}+\ldots$$

$$v_n=\frac{1}{n^2}$$

where v_n is n^{th} term of the auxiliary series Σv_n.

$$\frac{u_n}{v_n}=1+\frac{1}{8n^8}+\ldots$$

$$\therefore \quad \lim_{n\to\infty}\frac{u_n}{v_n}=1$$

which is a fixed finite non-zero quantity.

Hence comparison test can be applied.

But $\Sigma\frac{1}{n^2}$ is convergent as $p\,(=2)>1$ by p-test.

Hence Σu_n is also convergent by comparison test.

Sol. 36. Let the given series be denoted by Σu_n. Then

$$u_n=\left(n^3+1\right)^{1/3}-n$$

$$=n\left[\left(1+\frac{1}{n^3}\right)^{1/3}-1\right]$$

$$=n\left[1+\frac{1}{3}\frac{1}{n^3}+\frac{\frac{1}{3}\left(\frac{1}{3}-1\right)}{2!}\frac{1}{n^6}+\ldots-1\right]$$

$$=\frac{1}{3n^2}-\frac{1}{9}\frac{1}{n^5}+\ldots$$

$$v_n=\frac{1}{n^2}$$

where v_n is n^{th} term of the auxiliary series Σv_n.

$$\frac{u_n}{v_n} = \frac{1}{3} - \frac{1}{9n^3} + \ldots$$

$$\therefore \quad \underset{n \to \infty}{\text{lt.}} \frac{u_n}{v_n} = \frac{1}{3}$$

which is a fixed finite non-zero quantity.
Hence comparison test can be applied.

But $\Sigma \frac{1}{n^2}$ is convergent as $p(=2) > 1$ by p-test.

Hence Σu_n is also convergent by comparison test.

Sol. 37. Let the given series be denoted by Σu_n. Then

$$u_n = \frac{\sqrt{n^2+n+1} - \sqrt{n^2-n+1}}{n}$$

$$= \frac{\sqrt{n^2\left\{1 + \frac{n+1}{n^2}\right\}} - \sqrt{n^2\left\{1 - \frac{n-1}{n^2}\right\}}}{n}$$

$$= \left(1 + \frac{n+1}{n^2}\right)^{1/2} - \left(1 - \frac{n-1}{n^2}\right)^{1/2}$$

$$= \left[1 + \frac{1}{2}\frac{n+1}{n^2} + \frac{\frac{1}{2}\left(\frac{1}{2}-1\right)}{2!}\left(\frac{n+1}{n^2}\right)^2 + \ldots\right]$$

$$- \left[1 - \frac{1}{2}\frac{n-1}{n^2} + \frac{\frac{1}{2}\left(\frac{1}{2}-1\right)}{2!}\left(\frac{n-1}{n^2}\right)^2 - \ldots\right]$$

$$= \frac{(n+1)+(n-1)}{2n^2} - \frac{(n+1)^2 - (n-1)^2}{8n^4} - \ldots$$

$$= \frac{1}{n} - \frac{1}{2n^3} + \ldots$$

$$v_n = \frac{1}{n}$$

where v_n is the nth term of the auxiliary series Σv_n.

$$\frac{u_n}{v_n} = 1 - \frac{1}{2n^2} + \ldots$$

$$\therefore \quad \underset{n \to \infty}{\text{lt.}} \frac{u_n}{v_n} = 1$$

which is a fixed finite non-zero quantity,

Hence comparison test can be applied.

But $\Sigma \frac{1}{n}$ is divergent as $p = 1$ by p-test

Hence Σv_n is also divergent by comparison test.

Test the convergence of the following series :

Examples :

38. $\Sigma \frac{1}{\sqrt{n} + \sqrt{n-1}}$

39. $\Sigma \frac{n}{1 + n\sqrt{n+1}}$

40. $\Sigma \left[\frac{1}{n} - \log \frac{n+1}{n} \right]$

41. $\Sigma \sqrt{\frac{n}{2 + 3n^3}}$

42. $\Sigma \frac{1}{n^{1+1/n}}$.

Solutions :

Sol. 38. Let the given series be denoted by Σu_n. Then

$$u_n = \frac{1}{\sqrt{n} + \sqrt{n-1}} \times \frac{\sqrt{n} - \sqrt{n-1}}{\sqrt{n} - \sqrt{n-1}}$$

$$= \sqrt{n} - \sqrt{n-1}$$

Now proceeding as in Ex. 26, the series is divergent.

Sol. 39. Let the given series be denoted by Σu_n. Then

$$u_n = \frac{n}{1+n\sqrt{n+1}}$$

$$= \frac{n}{1+n\sqrt{n}\left(1+\frac{1}{n}\right)^{1/2}}$$

$$= \frac{n}{1+n^{3/2}\left[1+\frac{1}{2}\frac{1}{n}+\frac{\frac{1}{2}\left(\frac{1}{2}-1\right)}{2\,!}\frac{1}{n^2}+.....\right]}$$

$$= \frac{n}{1+n^{3/2}\left[1+\frac{1}{2n}-\frac{1}{8n^2}+.....\right]}$$

$$= \frac{n}{1+n^{3/2}+\frac{1}{2}n^{1/2}-\frac{1}{8n^{1/2}}+.....}$$

$$v_n = \frac{n}{n^{3/2}} = \frac{1}{n^{1/2}}$$

where v_n is n^{th} term of the auxiliary series Σv_n.

$$\frac{u_n}{v_n} = \frac{1}{1+\frac{1}{n^{3/2}}+\frac{1}{2n}-\frac{1}{8n^2}+.....}$$

$$\therefore \quad \lim_{n\to\infty} \frac{u_n}{v_n} = 1$$

which is a fixed finite non-zero quantity.

Hence comparison test can be applied.

But $\Sigma \frac{1}{n^{1/2}}$ is divergent as $p\left(=\frac{1}{2}\right)<1$ by p-test.

Hence Σu_n is also divergent by comparison test.

Sol. 40. Let the given series be denoted by Σu_n. Then

$$u_n = \frac{1}{n} - \log\left(\frac{n+1}{n}\right)$$

$$= \frac{1}{n} - \log\left(1 + \frac{1}{n}\right)$$

$$= \frac{1}{n} - \left(\frac{1}{n} - \frac{1}{2n^2} + \frac{1}{3n^3} + ...\right)$$

$$\left(\because \log(1+x) = x - \frac{x^2}{2} + \frac{x^3}{3} - \frac{x^4}{4} + ...\right)$$

$$= \frac{1}{2n^2} - \frac{1}{3n^3} + ...$$

$$v_n = \frac{1}{n^2}$$

where v_n is n^{th} term of the auxiliary series Σv_n.

$$\frac{u_n}{v_n} = \frac{1}{2} - \frac{1}{3n} + ...$$

$$\therefore \lim_{n\to\infty} \frac{u_n}{v_n} = \frac{1}{2}$$

which is a fixed finite non-zero quantity.

Hence comparison test can be applied.

But $\Sigma \frac{1}{n^2}$ is convergent as $p\,(= 2) > 1$ by p-test.

Hence Σu_n is also convergent by comparison test.

Sol. 41. Let the given series be divided by Σu_n. Then

$$u_n = \sqrt{\frac{n}{2+3n^3}}$$

$$v_n = \sqrt{\frac{n}{n^3}} = \frac{1}{n}$$

$$\therefore \qquad \frac{u_n}{v_n} = \sqrt{\frac{1}{\frac{2}{n^3}+3}}$$

$$\therefore \qquad \underset{n \to \infty}{\text{lt.}} \frac{u_n}{v_n} = \sqrt{\frac{1}{3}}$$

which is a fixed finite non-zero quantity.

Hence comparison test can be applied.

But $\Sigma \frac{1}{n}$ is divergent as $p = 1$ by p-test.

Hence Σu_n is also convergent by comparison test.

Sol. 42. Let the given series be divided by Σu_n. Then

$$u_n = \frac{1}{n^{1+1/n}}$$

Take $\quad v_n = \frac{1}{n}$

where v_n is n^{th} term of the auxiliary series Σv_n.

$$\frac{u_n}{v_n} = \frac{n}{n^{1+1/n}} = \frac{1}{n^{1/n}}$$

$$\therefore \qquad \underset{n \to \infty}{\text{lt.}} \frac{u_n}{v_n} = \underset{n \to \infty}{\text{lt}} = \frac{1}{n^{1/n}} = 1$$

which is a fixed finite non-zero quantity

$$\left(\begin{array}{ll} \because & \text{we know that} \\ & \underset{n\to\infty}{\text{lt}} \dfrac{\log n}{n} = 0 \\ \text{i.e.} & \underset{n\to\infty}{\text{lt}} \log (n)^{1/n} = 0 \\ \text{i.e.} & \underset{n\to\infty}{\text{lt}} n^{1/n} = e^0 = 1 \end{array} \right)$$

Hence comparison test can be applied.

But $\sum \frac{1}{n}$ is divergent as $p = 1$ by p-test.

Hence Σu_n is also divergent by comparison test.

Ex. 43. *Test the series*

$$\Sigma \sin \frac{1}{n}.$$

Sol. Let the given series be denoted by Σu_n. Then

$$u_n = \sin\left(\frac{1}{n}\right)$$

Take $\quad v_n = \frac{1}{n}$

where v_n is the nth term of the auxiliary series Σv_n.

$$\therefore \quad \frac{u_n}{v_n} = \frac{\sin\left(\frac{1}{n}\right)}{\frac{1}{n}}$$

$$\therefore \quad \underset{n \to \infty}{\text{lt.}} \frac{u_n}{v_n} = \underset{n \to \infty}{\text{lt.}} \frac{\sin\left(\frac{1}{n}\right)}{\frac{1}{n}}$$

$$= \underset{\frac{1}{n} \to 0}{\text{lt.}} \frac{\sin\left(\frac{1}{n}\right)}{\left(\frac{1}{n}\right)}$$

$$= 1,$$

$$\left(\because \underset{\theta \to 0}{\text{lt.}} \frac{\sin\theta}{\theta} = 1\right)$$

which is a fixed finite non-zero quantity.

Hence comparison test can be applied.

But $\Sigma \frac{1}{n}$ is divergent as $p = 1$ by p-test.

Hence Σu_n is also divergent by comparison test.

Ex. 44. *Test the series*

$$\Sigma \frac{1}{n} \sin \frac{1}{n}.$$

Sol. Let the given series be denoted by Σu_n. Then

$$u_n = \frac{1}{n} \sin \frac{1}{n}.$$

$$= \frac{1}{n}\left[\frac{1}{n} - \frac{1}{3!}\frac{1}{n^3} + \frac{1}{5!}\frac{1}{n^5} - .....\right]$$

$$= \frac{1}{n^2} - \frac{1}{6}\frac{1}{n^4} + \frac{1}{120}\frac{1}{n^6} - ...$$

$$v_n = \frac{1}{n^2}$$

where v_n is n^{th} term of the auxiliary series Σv_n.

$$\frac{u_n}{v_n} = 1 - \frac{1}{6n^2} + \frac{1}{120n^4} - ...$$

$$\therefore \quad \underset{n\to\infty}{\text{lt.}} \frac{u_n}{v_n} = 1$$

which is a fixed finite non-zero quantity.

Hence comparison test be applied.

But $\Sigma \frac{1}{n^2}$ is convergent as $p\ (= 2) > 1$ by p-test.

Hence Σu_n is also convergent by comparison test.

Ex. 45. *Test the series*

$$\Sigma \cos\left(\frac{1}{n}\right)$$

Sol. Let the term series be divided by Σu_n. Then

$$u_n = \cos\left(\frac{1}{n}\right)$$

$$= 1 - \frac{1}{2!n^2} + \frac{1}{4!}\frac{1}{n^4} ...$$

$$\left(\because \quad \cos\theta = 1 - \frac{\theta^2}{2!} + \frac{\theta^4}{4!} ...\right)$$

$$v_n = 1 = \frac{1}{n^0}$$

where v_n is n^{th} term of the auxiliary series Σv_n.

$$\therefore \qquad \frac{u_n}{v_n} = 1 - \frac{1}{2n^2} + \frac{1}{24n^4} \ldots$$

$$\therefore \qquad \underset{n \to \infty}{\text{lt.}} \frac{u_n}{v_n} = 1$$

which is a fixed finite non-zero quantity.

Hence comparison test can be applied.

But $\Sigma \frac{1}{n^0}$ is divergent as p (= 0) < 1 by p-test.

Hence Σu_n is also divergent by comparison test.

Ex. 46. *Test the series*

$$\Sigma \tan^{-1} \frac{1}{n}$$

Sol. Let the given series be divided by Σu_n. Then

$$u_n = \tan^{-1} \frac{1}{n}$$

$$= \frac{1}{n} - \frac{1}{3}\left(\frac{1}{n}\right)^3 + \frac{1}{5}\left(\frac{1}{n}\right)^5 - \ldots$$

$$\left[\because \ \tan^{-1} x = x - \frac{x^3}{3} + \frac{x^5}{5} \ldots . \to \text{Gregory's Series}\right]$$

$$\therefore \qquad v_n = \frac{1}{n}$$

where v_n is n^{th} term of the auxiliary series Σv_n.

$$\frac{u_n}{v_n} = 1 - \frac{1}{3n^2} + \frac{1}{5n^4} \ldots$$

$$\therefore \qquad \underset{n \to \infty}{\text{lt.}} \frac{u_n}{v_n} = 1$$

which is a fixed finite non-zero quantity,

Hence comparison test can be applied.

But $\Sigma\frac{1}{n}$ is divergent as $p = 1$ by p-test.

Hence Σu_n is also divergent by comparison test.

Ex. 47. *If* $a_n = \left(1+\frac{x}{n}\right)e^{-x/n}$

find whether the series whose n^{th} *term is* $a_n - 1$ *is convergent or divergent.*

Sol. Let the series where n^{th} term is $a_n - 1$ be denoted by Σu_n. Then,

$$u_n = a_n - 1$$

$$= \left(1+\frac{x}{n}\right)e^{-x/n} - 1$$

$$= \left(1+\frac{x}{n}\right)\left(1-\frac{x}{n}+\frac{x^2}{2n^2}-\frac{x^3}{6n^3}+\ldots\right)-1$$

$$= \left[1-\frac{x}{n}+\frac{x^2}{2n^2}-\frac{x^3}{6n^3}+\ldots+\frac{x}{n}-\frac{x^2}{n^2}+\frac{x^3}{2n^3}-\ldots\right]-1$$

$$= -\frac{x^2}{2n^2}+\frac{x^3}{3n^3}-\ldots$$

$$v_n = \frac{1}{n^2}$$

where v_n is n^{th} term of the auxiliary series Σv_n.

$$\frac{u_n}{v_n} = -\frac{x^2}{2}+\frac{x^3}{3n}-\ldots$$

$$\therefore \quad \underset{n\to\infty}{\text{lt.}} \frac{u_n}{v_n} = -\frac{x^2}{2}$$

which is a fixed finite non-zero quantity.

Hence comparison test can be applied.

But $\Sigma\frac{1}{n^2}$ is convergent as $p\ (=2) > 1$ by p-test.

Hence Σu_n is also convergent by comparison test.

1.17 Cauchy's Root Test Or Radical Test

Statement. An infinite series Σu_n of positive terms is convergent or divergent according as

$$\lim_{n \to \infty} (u_n)^{1/n} < 1 \text{ or } > 1.$$

If $\lim_{n \to \infty} (u)^{1/n} = 1$, the test fails.

Proof. Let $\Sigma u_n = u_1 + u_2 + u_3 + \ldots + u_n + \ldots$

be the given infinite series of positive terms.

Let us consider an auxiliary series

$$\Sigma v_n = v_1 + v_2 + v_3 + \ldots + v_n + \ldots$$
$$= r + r^2 + r^3 + \ldots + r^n + \ldots$$

which is clearly an infinite geometric progression with common ratio r. We know that Σv_n will be convergent if $r < 1$ and divergent if $r \geq 1$.

By comparison test, we know that if $\lim_{n \to \infty} \frac{u_n}{v_n}$ is a fixed finite non-zero quantity, then Σu_n and Σv_n either both converge or both diverge.

Let $\frac{u_n}{v_n} = k$

where k is a fixed finite non-zero quantity,

then $\frac{u_n}{r^n} = k$

or, $u_n = kr^n$

or, $(u_n)^{1/n} = k^{-1/n} \cdot r$

Then, $\lim_{n \to \infty} (u_n)^{1/n} = \lim_{n \to \infty} k^{1/n} \cdot r$

$= r,$

Now if $r < 1$, then

$$\underset{n \to \infty}{\text{lt.}} (u_n)^{1/n} < 1$$

and Σv_n will be convergent.

Hence Σu_n will also be covergent.

Again if $r > 1$, then

$$\underset{n \to \infty}{\text{lt.}} (u_n)^{1/n} > 1$$

and Σv_n will be divergent,

Hence Σu_n will also be divergent.

If $r = 1$, then

$$\underset{n \to \infty}{\text{lt.}} (u_n)^{1/n} = 1$$

i.e. r may be infinitesimally less or greater than 1, so nothing can be said about the nature of the series in this case.

Hence this test fails when $r = 1$

i.e. when $\underset{n \to \infty}{\text{lt.}} (u_n)^{1/n} = 1$

Illustrative Examples

Ex. 1. *Test the series*

$$\Sigma \left(1 + \frac{1}{n}\right)^{n^2}.$$

Sol. Let the given series be denoted by Σu_n. Then,

$$u_n = \left(1 + \frac{1}{n}\right)^{n^2}$$

$$(u_n)^{1/n} = \left(1 + \frac{1}{n}\right)^{n}$$

$$\therefore \underset{n \to \infty}{\text{lt.}} (u_n)^{1/n} = \underset{n \to \infty}{\text{lt.}} \left(1 + \frac{1}{n}\right)^{n}$$

$$= e > 1$$

Hence by Cauchy's root test, Σu_n is divergent.

Ex. 2. *Test the series*

$$\Sigma\left(1+\frac{1}{n}\right)^{-n^2}$$

Sol. Let the given series be denoted by Σu_n. Then,

$$u_n = \left(1+\frac{1}{n}\right)^{-n^2}$$

$$(u_n)^{1/n} = \left(1+\frac{1}{n}\right)^{-n} = \frac{1}{\left(1+\frac{1}{n}\right)^n}$$

$$\therefore \quad \underset{n\to\infty}{\text{lt.}} (u_n)^{1/n} = \frac{1}{e} < 1$$

Hence by Cauchy's root test, the given series is convergent.

Ex. 3. *Test the series*

$$\Sigma\left(1-\frac{1}{n}\right)^{n^2}.$$

Sol. Here $\quad u_n = \left(1-\frac{1}{n}\right)^{n^2}$

$$\therefore \quad (u_n)^{1/n} = \left(1-\frac{1}{n}\right)^n = \left[\left(1-\frac{1}{n}\right)^{-n}\right]^{-1}$$

$$\therefore \quad \underset{n\to\infty}{\text{lt.}} (u_n)^{1/n} = e^{-1} = \frac{1}{e} < 1$$

Hence by Cauchy's root test, the given series is convergent.

Ex. 4. *Test the series*

$$\Sigma\left\{\log\left(1+\frac{1}{n}\right)\right\}^n.$$

Sol. Here, $\quad u_n = \left\{\log\left(1+\frac{1}{n}\right)\right\}^n$

$$(u_n)^{1/n} = \log\left(1+\frac{1}{n}\right)$$

$$\therefore \quad \underset{n\to\infty}{\text{lt.}} (u_n)^{1/n} = \log 1 = 0 < 1$$

Hence by Cauchy's root test the given series is convergent.

Ex. 5. *Test the convergence of the series*

$$1 + \frac{x}{2} + \frac{x^2}{3^2} + \frac{x^3}{4^3} + \ldots.$$

Sol. Neglecting the first term,

let the given series be denoted by Σu_n. Then,

$$u_n = \frac{x^n}{(n+1)^n}$$

$$\therefore \quad (u_n)^{1/n} = \frac{x}{n+1}$$

$$\therefore \quad \underset{n\to\infty}{\text{lt.}} (u_n)^{1/n} = \underset{n\to\infty}{\text{lt.}} \frac{x}{n+1}$$

$$= 0 < 1 \text{ for all values of } x.$$

Hence by Cauchy's root test, the given series is convergent.

Ex. 6. *Test the series*

$$\frac{1}{2} + \frac{2}{3}x + \left(\frac{3}{4}\right)^2 x^2 + \left(\frac{4}{5}\right)^3 x^3 + \ldots$$

Sol. Neglecting the first term,

let the given series be denoted by Σu_n. Then,

$$u_n = \left(\frac{n+1}{n+2}\right)^n x^n$$

$$\therefore \quad (u_n)^{1/n} = \frac{n+1}{n+2}x$$

$$= \frac{1+\frac{1}{n}}{1+\frac{2}{n}}x$$

$$\therefore \lim_{n\to\infty} (u_n)^{1/n} = x$$

Hence by Cauchy's root test the given series will be convergent or divergent according as $x < 1$ or > 1.

When $x = 1$,

$$u_n = \left(\frac{n+1}{n+2}\right)^n$$

$$v_n = 1 = \frac{1}{n^0}$$

where v_n is n^{th} term of auxiliary series Σv_n.

$$\therefore \frac{u_n}{v_n} = \left(\frac{n+1}{n+2}\right)^n$$

$$= \frac{\left(1+\frac{1}{n}\right)^n}{\left(1+\frac{2}{n}\right)^n}$$

$$\therefore \lim_{n\to\infty} \frac{u_n}{v_n} = \lim_{n\to\infty} \frac{\left(1+\frac{1}{n}\right)^n}{\left(1+\frac{2}{n}\right)^n}$$

$$= \frac{e}{e^2} = \frac{1}{e}$$

which is a fixed finite non-zero quantity.

Hence comparison test can be applied.

But $\Sigma \frac{1}{n^0}$ is divergent as $p\ (= 0) < 1$ by p-test.

Hence Σu_n is also divergent by comparison test.

Ex. 7. *Test the series*

$$\Sigma (n^{1/n} - 1)^n$$

Sol. Here,. $u_n = (n^{1/n} - 1)^n$

$$(u_n)^{1/n} = n^{1/n} - 1$$

$$\therefore \lim_{n\to\infty} (u_n)^{\frac{1}{n}} = \lim_{n\to\infty} (n^{1/n} - 1)$$

$$\left| \because \lim_{n\to\infty} n^{\frac{1}{n}} = 1 \right.$$

Hence by Cauchy's root test the given series is convergent.

Ex. 8. *Test the series*

$$\sum_{n=2}^{\infty} \left(\frac{1}{\log n}\right)^n$$

Sol. Let u_n be the general term of the given series, then

$$u_n = \left(\frac{1}{\log n}\right)^n$$

$$(u_n)^{1/n} = \frac{1}{\log n}$$

$$\therefore \lim_{n\to\infty} (u_n)^{1/n} = \frac{1}{\log \infty} = \frac{1}{\infty} = 0 < 1$$

Hence by Cauchy's root test, the given series is convergent.

Ex. 9. *Test the series*

$$\left(\frac{2^2}{1^2} - \frac{2}{1}\right)^{-1} + \left(\frac{3^3}{2^3} - \frac{3}{2}\right)^{-2} + \left(\frac{4^4}{3^4} + \frac{4}{3}\right)^{-3} + \ldots$$

Sol. Let the given series be denoted by Σu_n. Then

$$u_n = \left[\frac{(n+1)^{n+1}}{n^{n+1}} - \frac{n+1}{n}\right]^{-n}$$

$$= \left[\left(\frac{n+1}{n}\right)^{n+1} - \frac{n+1}{n}\right]^{-n}$$

$$(u_n)^{1/n} = \left[\left(\frac{n+1}{n}\right)^{n+1} - \frac{n+1}{n}\right]^{-1}$$

$$= \left(\frac{n+1}{n}\right)^{-1}\left[\left(\frac{n+1}{n}\right)^{n}-1\right]^{-1}$$

$$= \left(1+\frac{1}{n}\right)^{-1}\left[\left(1+\frac{1}{n}\right)^{n}-1\right]^{-1}$$

$$\therefore \quad \underset{n\to\infty}{\text{lt.}} (u_n)^{1/n} = (1+0)^{-1}(e-1)^{-1}$$

$$= \frac{1}{e-1} < 1. \qquad (\because e = 2.718)$$

Hence by Cauchy's root test, the given series is convergent.

Ex. 10. *Test the series*

$$\Sigma\left(1+\frac{1}{\sqrt{n}}\right)^{-n^{\frac{3}{2}}}$$

Sol. Let the given series be denoted by Σu_n.
Then,

$$u_n = \left(1+\frac{1}{\sqrt{n}}\right)^{-n^{3/2}}$$

$$\therefore \quad (u_n)^{1/n} = \left[\left(1+\frac{1}{\sqrt{n}}\right)^{-n^{3/2}}\right]^{1/n}$$

$$= \left(1+\frac{1}{\sqrt{n}}\right)^{-\sqrt{n}}$$

$$= \frac{1}{\left(1+\frac{1}{\sqrt{n}}\right)^{\sqrt{n}}}$$

$$\therefore \quad \underset{n\to\infty}{\text{lt.}} (u_n)^{1/n} = \underset{n\to\infty}{\text{lt.}} \frac{1}{\left(1+\frac{1}{\sqrt{n}}\right)^{\sqrt{n}}} \quad \left(\begin{array}{l}\because \textit{If } n\to\infty, \\ \text{then } \sqrt{n}\to\infty\end{array}\right)$$

$$= \frac{1}{e} < 1 \qquad | \because e = 2.718$$

Hence by Cauchy's root test, the given series is convergent.

Ex. 11. *Test the series*

$$\Sigma \frac{n^{n^2}}{(n+1)^{n^2}}$$

Sol. Here,

$$u_n = \frac{n^{n^2}}{(n+1)^{n^2}} = \left(\frac{n}{n+1}\right)^{n^2}.$$

$$(u_n)^{1/n} = \left[\left(\frac{n}{n+1}\right)^{n^2}\right]^{1/n} = \left(\frac{n}{n+1}\right)^n$$

$$= \frac{1}{\left(\frac{n+1}{n}\right)^n}$$

$$\therefore \quad \underset{n \to \infty}{\text{lt.}} (u_n)^{1/n} = \frac{1}{e} < 1 \qquad | \because e = 2.718$$

Hence by Couchy's root test, the given series is convergent.

Ex. 12. *Test the series*

$$\Sigma \frac{n^{n^2}}{\left(n + \frac{1}{4}\right)^{n^2}}$$

Sol. Here, $u_n = \left(\dfrac{n}{n + \frac{1}{4}}\right)^{n^2}$

$$\therefore \quad (u_n)^{1/n} = \left(\frac{n}{n + \frac{1}{4}}\right)^n$$

$$= \frac{1}{\left[\left(1+\frac{1}{4n}\right)^{4n}\right]^{1/4}}$$

$$\therefore \quad \underset{n\to\infty}{\text{lt.}} (u_n)^{1/n} = \frac{1}{e^{1/4}}$$

$$\left(\because \begin{array}{c} n\to\infty, \\ \text{then } n\to\infty \end{array}\right)$$

$$< 1 \quad | \because e^{1/4} > 1$$

Hence by Cauchy's root test, the given series is convergent.

Ex. 13. *Test the series*

$$\Sigma \frac{(n+1)^n x^n}{n^{n+1}}.$$

Sol. Let the given series be denoted by Σu_n.

$$\text{Then,} \quad u_n = \frac{(n+1)^n x^n}{n^{n+1}}$$

$$= \left[\frac{(n+1)x}{n}\right]^n \frac{1}{n}$$

$$\therefore \quad (u_n)^{1/n} = \frac{(n+1)x}{n} \frac{1}{n^{1/n}}$$

$$= \left(1+\frac{1}{n}\right) x \frac{1}{n^{1/n}}$$

$$\therefore \quad \underset{n\to\infty}{\text{lt.}} (u_n)^{1/n} = x$$

Hence Σu_n is convergent or divergent according as $x < 1$ or >1.
Again, if $x = 1$, then

$$u_n = \frac{(n+1)^n}{n^{n+1}}$$

$$v_n = \frac{n^n}{n^{n+1}} = \frac{1}{n}$$

where v_n is n^{th} term of the auxiliary series Σv_n.

$$\frac{u_n}{v_n} = \frac{(n+1)^n}{n^n} = \left(1 + \frac{1}{n}\right)^n$$

$$\therefore \quad \underset{n \to \infty}{\text{lt.}} \frac{u_n}{v_n} = e$$

which is a fixed finite non-zero quantity.

Hence comparison test can be applied.

But $\Sigma \frac{1}{n}$ is divergent as $p = 1$ by p-test.

Hence Σu_n is also divergent by comparison test.

Ex. 14. *Test the series*

$$\Sigma \left(\frac{1 + nx}{n}\right)^n.$$

Sol. Here,
$$u_n = \left(\frac{1 + nx}{n}\right)^n$$

$$(u_n)^{1/n} = \frac{1 + nx}{n}$$

$$\therefore \quad \underset{n \to \infty}{\text{lt.}} (u_n)^{1/n} = \underset{n \to \infty}{\text{lt.}} \left(\frac{1 + nx}{n}\right)$$

$$= \underset{n \to \infty}{\text{lt.}} \left(\frac{1}{n} + x\right)$$

$$= x$$

Hence by Cauchy's root test the given series is convergent or divergent according as

$$x < 1 \text{ or } > 1,$$

If $x = 1$,

then,
$$u_n = \left(\frac{n+1}{n}\right)^n$$

$$v_n = 1 = \frac{1}{n^0}$$

where v_n is the nth term of the auxiliary series Σv_n.

$$\frac{u_n}{v_n} = \left(1 + \frac{1}{n}\right)^n$$

$$\therefore \quad \underset{n\to\infty}{\text{lt.}} \frac{u_n}{v_n} = \underset{n\to\infty}{\text{lt.}} \left(1+\frac{1}{n}\right)^n = e$$

which is a fixed finite non-zero quantity.

Hence comparison test can be applied.

But $\Sigma \frac{1}{n^0}$ is divergent as p (= 0) < 1 by p-test.

Hence Σu_n is also divergent by comparison test.

Ex. 15. *Test for convergence the series*

$$a + b + a^2 + b^2 + a^3 + b^3 + \ldots$$

Sol. The given series can be written as

$$(a + a^2 + a^3 + \ldots) + (b + b^2 + b^3 + \ldots)$$

$$= \Sigma u_n + \Sigma v_n$$

where $\quad \Sigma u_n = a + a^2 + a^3 + \ldots$

and $\quad \Sigma v_n = b + b^2 + b^3 + \ldots$

Now consider the series Σu_n.

$$u_n = a^n$$

$$(u_n)^{1/n} = a$$

$$\therefore \quad \underset{n\to\infty}{\text{lt.}} (u_n)^{1/n} = a$$

Hence by Cauchy's root test, Σu_n will be convergent or divergent according as

$$a < 1 \text{ or } > 1,$$

Now if $\quad a = 1$, Σu_n becomes

$$1 + 1 + 1 + 1 + \ldots$$

$$\therefore \quad S_n = n$$

$$\therefore \lim_{n\to\infty} S_n = \infty$$

Hence Σu_n is divergent.

Thus Σu_n is convergent when $a < 1$

and divergent when $a \geq 1$

Similarly we can say that Σv_n is

convergent when $b < 1$

and divergent when $b \geq 1$

Hence the given series will be convergent only when $0 < a < 1$ and $0 < b < 1$ and divergent otherwise, since a and b are (+)ve numbers.

1.18 D'Alembert's Test

Or

Ratio Test

Statement. An infinite series Σu_n of positive terms is convergent, if from and after some fixed terms,

$$\frac{u_{n+1}}{u_n} < k < 1$$

where k is fixed number and is divergent if $\frac{u_{n+1}}{u_n} > 1$.

Proof.

Let the given series after leaving a finite number of terms, if necessary, be denoted by $\Sigma u_n = u_1 + u_2 + u_3 + \ldots$

and let $\frac{u_2}{u_1} < k$,

$$\frac{u_3}{u_2} < k,$$

$$\frac{u_4}{u_3} < k$$

$$\vdots$$

and so on

where $k < 1$

Now, $u_1 + u_2 + u_3 + u_4 + \ldots$

$$= u_1\left[1+\frac{u_2}{u_1}+\frac{u_3}{u_2}.\frac{u_2}{u_1}+\frac{u_4}{u_3}.\frac{u_3}{u_2}.\frac{u_2}{u_1}+\ldots\right]$$

$$< u_1\left[1+k+k^2+k^3+\ldots.\right]$$

$$< \frac{u_1}{1-k} \qquad (as\ k<1)$$

Hence the given series is convergent.

Again, let $\frac{u_2}{u_1}>1, \frac{u_3}{u_2}>1, \frac{u_4}{u_3}>1,\ldots$ and so on

Then $u_2 > u_1$

$u_3 > u_2$

$u_4 > u_3$ ...... and so on

i.e. $S_n = u_1 + u_2 + u_3 + \ldots\ldots\ldots + u_n > nu_1$

$$\therefore \quad \underset{n\to\infty}{\text{lt.}} S_n > \underset{n\to\infty}{\text{lt.}} nu_1$$

$$\text{i.e} \quad \underset{n\to\infty}{\text{lt.}} S_n \to \infty$$

Hence the series Σu_n is divergent.

When $\underset{n\to\infty}{\text{lt.}} \frac{u_{n+1}}{u_n} = 1$ then it means that $\frac{u_{n+1}}{u_n}$ may be infinitesimally less or greater than 1, so nothing can be said about the nature of the series in this case, hence this test fails when

$$\underset{n\to\infty}{\text{lt.}} \frac{u_{n+1}}{u_n} = 1$$

and we determine the convergence or divergence of the given series by some other suitable test in this case.

Note : While testing the convergence or divergence of the series, D'Alembert's ratio test is generally applied in its following form. The series Σu_n of positive terms is convergent or divergent according as

$$\underset{n\to\infty}{\text{lt.}} \frac{u_n}{u_{n+1}} \text{ is } > 1 \text{ or } < 1.$$

If this limits is equal to 1, the test fails.

Illustrative Examples

Ex. 1. *Discuss the convergence of*

$$\frac{1}{1+2}+\frac{2}{1+2^2}+\frac{3}{1+2^3}+\ldots$$

Sol. Here, $u_n = \dfrac{n}{1+2^n}$

$$\therefore \quad u_{n+1} = \frac{n+1}{1+2^{n+1}} \quad \text{(replace } n \text{ by } n+1\text{)}$$

$$\frac{u_n}{u_{n+1}} = \frac{n}{n+1}\,\frac{1+2^{n+1}}{1+2^n}$$

$$= \frac{1}{1+\frac{1}{n}}\,\frac{2^{n+1}\left(1+\frac{1}{2^{n+1}}\right)}{2^n\left(1+\frac{1}{2^n}\right)}$$

$$= \frac{2}{1+\frac{1}{n}}\,\frac{1+\frac{1}{2^{n+1}}}{1+\frac{1}{2^n}}$$

$$\therefore \quad \lim_{n\to\infty} \frac{u_n}{u_{n+1}} = 2 \text{ which is} > 1$$

Hence by ratio test, the given series is convergent.

Ex. 2. *Show that the series*

$$1+\frac{2^p}{2!}+\frac{3^p}{3!}+\frac{4^p}{4!}+\ldots$$

is convergent for all values of p.

Sol. Here, $u_n = \dfrac{n^p}{n!}$

$$\therefore \quad u_{n+1} = \frac{(n+1)^p}{n+1!}$$

$$\frac{u_n}{u_{n+1}} = \frac{n^p}{(n+1)^p} \frac{n+1!}{n!}$$

$$= \frac{n+1}{\left(1+\frac{1}{n}\right)^p}$$

$$\therefore \quad \underset{n\to\infty}{\text{lt.}} \frac{u_n}{u_{n+1}} = \infty$$

which is >1.

Hence by ratio test, the given series is convergent.

Ex. 3. *Test the series*

$$\frac{2\,!}{3} + \frac{3\,!}{3^2} + \frac{4\,!}{3^3} + \ldots$$

Sol. Here, $$u_n = \frac{n+1\,!}{3^n}$$

$$\therefore \quad u_{n+1} = \frac{n+2\,!}{3^{n+1}}$$

$$\therefore \quad \frac{u_n}{u_{n+1}} = \frac{n+1\,!}{n+2\,!} \frac{3^{n+1}}{3^n}$$

$$= \frac{3}{n+2}$$

$$\therefore \quad \underset{n\to\infty}{\text{lt.}} \frac{u_n}{u_{n+1}} = 0$$

which is <1.

Hence by ratio test, the given series is divergent.

Ex. 4. *Test the series*

$$1 + \frac{2^2}{2!} + \frac{3^2}{3!} + \frac{4^2}{4!} + \ldots$$

Sol. Here, $$u_n = \frac{n^2}{n\,!}$$

$$\therefore \qquad u_{n+1} = \frac{(n+1)^2}{n+1\,!}$$

$$\frac{u_n}{u_{n+1}} = \frac{n^2}{(n+1)^2}\frac{n+1\,!}{n\,!}$$

$$= \frac{1}{\left(1+\frac{1}{n}\right)^2}(n+1)$$

$$\therefore \qquad \underset{n\to\infty}{\text{lt.}} \frac{u_n}{u_{n+1}} = \infty > 1$$

Hence by ratio test, the given series is convergent.

Ex. 5. *Test the series*

$$\frac{2}{1^2+1} + \frac{2^2}{2^2+1} + \frac{2^3}{3^2+1} + \ldots$$

Sol. Here, $\quad u_n = \dfrac{2^n}{n^2+1}$

$$\therefore \qquad u_{n+1} = \frac{2^{n+1}}{(n+1)^2+1} = \frac{2^{n+1}}{n^2+2n+2}$$

$$\frac{u_n}{u_{n+1}} = \frac{2^n}{n^2+1} \cdot \frac{n^2+2n+2}{2^{n+1}}$$

$$= \frac{1+\frac{2}{n}+\frac{2}{n^2}}{1+\frac{1}{n^2}} \cdot \frac{1}{2}$$

$$\therefore \qquad \underset{n\to\infty}{\text{lt.}} \frac{u_n}{u_{n+1}} = \frac{1}{2} < 1$$

Hence by ratio test, the given series is divergent.

Ex. 6. *Test the series*

$$\frac{1\,!}{5} + \frac{2\,!}{5^2} + \frac{3\,!}{5^3} + \ldots$$

Sol. Here, $u_n = \frac{n!}{5^n}$

$$u_{n+1} = \frac{n+1!}{5^{n+1}}$$

$$\therefore \quad \frac{u_n}{u_{n+1}} = \frac{n!}{n+1!} \cdot \frac{5^{n+1}}{5} = \frac{5}{n+1}$$

$$\therefore \quad \underset{n\to\infty}{\text{lt.}} \frac{u_n}{u_{n+1}} = \frac{5}{\infty+1} = \frac{5}{\infty} = 0 < 1$$

Hence by ratio test, the given series is divergent.

Ex. 7. *Test the series*

$$1 + 2x + 3x^2 + 4x^3 + \ldots$$

Sol. Here, $u_n = nx^{n-1}$

$$u_{n+1} = (n+1)x^n$$

$$\frac{u_n}{u_{n+1}} = \frac{n}{n+1} \frac{x^{n-1}}{x^n} = \frac{n}{n+1} \cdot \frac{1}{x}$$

or,
$$\frac{u_n}{u_{n+1}} = \frac{1}{1+\frac{1}{n}} \frac{1}{x}$$

$$\therefore \quad \underset{n\to\infty}{\text{lt.}} \frac{u_n}{u_{n+1}} = \frac{1}{x}$$

$\therefore$ By ratio test the given series is

convergent if $\frac{1}{x} > 1$ i.e. $x < 1$

divergent if $\frac{1}{x} < 1$ i.e. $x > 1$

and the test fails if $\frac{1}{x} = 1$ i.e. $x = 1$

Now when $x = 1$, the series becomes

$$1 + 2 + 3 + 4 + \ldots.$$

S_n = Sum of n terms of the series

$= 1 + 2 + 3 + \ldots. + n$

$$= \frac{n(n+1)}{2}$$

$$\therefore \lim_{n \to \infty} S_n = \infty$$

Hence the series is divergent.

Ex. 8. *Test the series*

$1 + 3x + 5x^2 + 7x^3 + ...$

Sol. Here $u_n = (2n-1)x^{n-1}$

$\therefore \quad u_{n+1} = (2n+1)x^n$

$$\therefore \quad \frac{u_n}{u_{n+1}} = \frac{2n-1}{2n+1} \cdot \frac{x^{n-1}}{x^n}$$

$$= \frac{2 - \frac{1}{n}}{2 + \frac{1}{n}} \cdot \frac{1}{x}$$

$$\therefore \lim_{n \to \infty} \frac{u_n}{u_{n+1}} = \frac{1}{x}$$

Hence by ratio test the given series is

convergent if $\frac{1}{x} > 1$ i.e. $x < 1$,

divergent if $\frac{1}{x} < 1$ i.e. $x > 1$,

and the test fails if $\frac{1}{x} = 1$ i.e. $x = 1$.

When $x = 1$, the series becomes

$1 + 3 + 5 + 7 + ...$

Since each term is +ve and greater than a fixed finite quantity $\frac{1}{2}$.

Hence the given series is divergent.

Ex. 9. *Test the series*

$$1+\frac{x}{2}+\frac{x^2}{5}+\frac{x^3}{10}+\ldots\frac{x^n}{n^2+1}+\ldots$$

Sol. After leaving the first term which does not affect the nature of the series,

let the given series be denoted by Σu_n. Then

$$u_n = \frac{x^n}{n^2+1}$$

$$u_{n+1} = \frac{x^{n+1}}{(n+1)^2+1} = \frac{x^{n+1}}{n^2+2n+2}$$

$$\frac{u_n}{u_{n+1}} = \frac{x^n}{x^n+1} \cdot \frac{n^2+2n+2}{n^2+1}$$

$$= \frac{1+\frac{2}{n}+\frac{2}{n^2}}{1+\frac{1}{n^2}} \cdot \frac{1}{x}$$

$$\therefore \quad \lim_{n\to\infty} \frac{u_n}{u_{n+1}} = \frac{1}{x}$$

Hence by ratio test, the given series is

convergent if $\frac{1}{x} > 1$ i.e. $x < 1$,

divergent if $\frac{1}{x} < 1$ i.e. $x > 1$

and the test fails if $\frac{1}{x} = 1$ i.e. $x = 1$

Now when $x = 1$,

$$u_n = \frac{1}{n^2+1}$$

$$v_n = \frac{1}{n^2+1}$$

when v_n is n^{th} term of the auxiliary series Σv_n,

$$\frac{u_n}{v_n} = \frac{n^2}{n^2+1}$$

$$\frac{u_n}{v_n} = \frac{1}{1+\frac{1}{n^2}}$$

$$\therefore \quad \underset{n\to\infty}{\text{lt.}} \frac{u_n}{v_n} = 1$$

which is a fixed finite non-zero quantity.

Hence comparison test can be applied.

But $\Sigma \frac{1}{n^2}$ is convergent as $p\ (= 2) > 1$ by p-test.

Hence by comparison test, the given series is also convergent.

Ex. 10. *Test the series whose* n^{th} *term is*

$$\frac{n+1}{n^2} x^n.$$

Sol. Here,

$$u_n = \frac{n+1}{n^2} x^n$$

$$u_{n+1} = \frac{n+1+1}{(n+1)^2} x^{n+1}$$

$$= \frac{n+2}{n^2+2n+1} x^{n+1}$$

$$\frac{u_n}{u_{n+1}} = \frac{n+1}{n+2} \cdot \frac{n^2+2n+1}{n^2} \frac{x^n}{x^{n+1}}$$

$$= \frac{1+\frac{1}{n}}{1+\frac{2}{n}} \cdot \left(1 + \frac{2}{n} + \frac{1}{n^2}\right) \frac{1}{x}$$

$$\therefore \quad \underset{n\to\infty}{\text{lt.}} \frac{u_n}{u_{n+1}} = \frac{1}{x}$$

Hence by ratio test the given series is

convergent if $\frac{1}{x} > 1$ i.e. $x < 1$

divergent if $\frac{1}{x} < 1$ i.e. $x < 1$

and the test fails if $\frac{1}{x} = 1$ i.e. $x = 1$

When $x = 1$,

$$u_n = \frac{n+1}{n^2}$$

$$v_n = \frac{n}{n^2} = \frac{1}{n}$$

where v_n is n^{th} term of the auxiliary series Σv_n.

$$\frac{u_n}{v_n} = \frac{n+1}{n} = 1 + \frac{1}{n}$$

$$\therefore \quad \underset{n \to \infty}{\text{lt.}} \frac{u_n}{v_n} = 1$$

which is a fixed finite non-zero quantity,

Hence comparison test can be applied.

But $\Sigma \frac{1}{n}$ is divergent as $p = 1$ by p-test.

Hence the given series is also divergent.

Ex. 11. *Test the series*

$$2x + \frac{3x^2}{8} + \frac{4x^3}{27} + \ldots + \frac{(n+1)x^n}{n^3} + \ldots$$

Sol. Here, $$u_n = \frac{(n+1)x^n}{n^3}$$

$$u_{n+1} = \frac{(n+2)x^{n+1}}{(n+1)^3}$$

$$\frac{u_n}{u_{n+1}} = \frac{n+1}{n+2}\left(\frac{n+1}{n}\right)^3 \frac{x^n}{x^{n+1}}$$

$$=\frac{1+\frac{1}{n}}{1+\frac{2}{n}}\left(1+\frac{1}{n}\right)^3\frac{1}{x}$$

$$\therefore \quad \lim_{n\to\infty}\frac{u_n}{v_{n+1}}=\frac{1}{x}$$

Hence by ratio test the given series is

convergent if $\frac{1}{x}>1$ i.e. $x<1$

divergent if $\frac{1}{x}<1$ i.e. $x>1$

and the test fails if $\frac{1}{x}=1$ i.e. $x=1$

When $\quad x=1$,

$$u_n=\frac{n+1}{n^3}$$

$$v_n=\frac{n}{n^3}=\frac{1}{n^2}$$

where v_n is n^{th} term of the auxiliary series Σv_n.

$$\frac{u_n}{v_n}=\frac{n+1}{n}=1+\frac{1}{n}$$

$$\therefore \quad \lim_{n\to\infty}\frac{u_n}{v_{n+1}}=1$$

which is a fixed finite non-zero quantity.

Hence comparison test can be applied.

But $\Sigma\frac{1}{n^2}$ is convergent as $p\ (=2)>1$ by p-test.

Hence the given series is also convergent by comparison test.

Ex. 12. *Test the series*

$$1+\frac{x^2}{2}+\frac{x^4}{4}+\frac{x^6}{6}+\ldots(x>0)$$

Sol. Leave the first term and then let the given series be denoted by Σu_n. Then

$$u_n = \frac{x^{2n}}{2n}$$

$$\therefore \quad u_{n+1} = \frac{x^{2n+2}}{2n+2}$$

$$\frac{u_n}{u_{n+1}} = \frac{x^{2n}}{x^{2n+2}} \cdot \frac{2n+2}{2n}$$

$$= \left(1 + \frac{1}{n}\right)\frac{1}{x^2}$$

$$\therefore \quad \underset{n\to\infty}{\text{lt.}} \frac{u_n}{u_{n+1}} = 1$$

Hence by ratio test the given series is

convergent if $\frac{1}{x^2} > 1$ i.e. $x^2 < 1$

divergent if $\frac{1}{x^2} < 1$ i.e. $x^2 > 1$

and the test fails if $\frac{1}{x^2} = 1$ i.e. $x^2 = 1$

When $\quad x^2 = 1$

$$u_n = \frac{1}{2n}$$

$$v_n = \frac{1}{2n}$$

where v_n is n^{th} term of the auxiliary series Σv_n.

$$\frac{u_n}{v_n} = \frac{1}{2}$$

$$\therefore \quad \underset{n\to\infty}{\text{lt.}} \frac{u_n}{v_n} = \frac{1}{2}$$

which is a fixed finite non-zero quantity.

Hence comparison test can be applied.

But $\Sigma\frac{1}{n}$ is divergent as $p = 1$ by p-test.

Hence the given series is also divergent by comparison test.

Ex. 13. *Test the series*

$$x + \frac{3x^2}{5} + \frac{8}{10}x^3 + \frac{15}{17}x^4 + \dots + \frac{n^2-1}{n^2+1}x^n + \dots$$

Sol. Here
$$u_n = \frac{n^2-1}{n^2+1}.x^n$$

$$u_{n+1} = \frac{(n+1)^2-1}{(n+1)^2+1}x^{n+1}$$

$$= \frac{n^2+2n}{n^2+2n+2}x^{n+1}$$

$$\frac{u_n}{u_{n+1}} = \frac{n^2-1}{n^2+1}.\frac{n^2+2n+2}{n^2+2n}\frac{x^n}{x^{n+1}}$$

$$= \frac{1-\frac{1}{n^2}}{1+\frac{1}{n^2}}\ \frac{1+\frac{2}{n}+\frac{2}{n^2}}{1+\frac{2}{n}}\ \frac{1}{x}$$

$$\therefore \quad \underset{n\to\infty}{\text{lt.}} \frac{u_n}{u_{n+1}} = \frac{1}{x}$$

Hence by ratio test, the given series is

convergent if $\frac{1}{x} > 1$ i.e. $x < 1$

divergent if $\frac{1}{x} < 1$ i.e. $x > 1$

and the test fails if $\frac{1}{x} = 1$ i.e. $x = 1$

When $x = 1$,

$$u_n = \frac{n^2 - 1}{n^2 + 1}$$

$$v_n = \frac{n^2}{n^2} = 1 = \frac{1}{n^0}$$

where v_n is the nth term of the auxiliary series Σv_n.

$$\frac{u_n}{v_n} = \frac{n^2 - 1}{n^2 + 1} = \frac{1 - \frac{1}{n^2}}{1 + \frac{1}{n^2}}$$

$$\therefore \quad \underset{n \to \infty}{\text{lt.}} \frac{u_n}{v_n} = 1$$

which is a fixed finite non-zero quantity.

Hence comparison test can be applied.

But $\Sigma \frac{1}{n^0}$ is divergent as $p(= 0) < 1$ by p-test.

Hence the given series is also divergent by comparison test.

Ex. 14. *Test for convergence or divergence the series whose n^{th} term is $\left(\sqrt{n^2 + 1} - n\right)x^{2n}$.*

Sol. Here,

$$u_n = \left(\sqrt{n^2 + 1} - n\right)x^{2n}$$

$$u_{n+1} = \left[\sqrt{(n+1)^2 + 1} - (n+1)\right]x^{2n+2}$$

$$= \left[\sqrt{n^2 + 2n + 2} - (n+1)\right]x^{2n+2}$$

$$\frac{u_n}{u_{n+1}} = \frac{\sqrt{n^2 + 1} - n}{\sqrt{n^2 + 2n + 2} - (n+1)} \frac{x^{2n}}{x^{2n+2}}$$

$$= \frac{\sqrt{n^2 + 1} - n}{\sqrt{n^2 + 2n + 2} - (n+1)} \frac{1}{x^2}$$

$$= \frac{n\left[\left(1+\frac{1}{n^2}\right)^{1/2}-1\right]}{n\left[\left(1+\frac{2}{n}+\frac{2}{n^2}\right)^{1/2}-\left(1+\frac{1}{n}\right)\right]}\frac{1}{x^2}$$

$$= \frac{1+\frac{1}{2}\frac{1}{n^2}-\frac{1}{8n^4}+\ldots-1}{1+\frac{1}{n}+\frac{1}{n^2}-\frac{1}{8}\left(\frac{2}{n}+\frac{2}{n^2}\right)^2+\ldots-1-\frac{1}{n}}\frac{1}{x^2}$$

$$= \frac{\frac{1}{2n^2}-\frac{1}{8n^4}+\ldots}{\frac{1}{2n^2}-\frac{1}{2n^3}+\ldots}\frac{1}{x^2}$$

$$= \frac{1-\frac{1}{4n^2}+\ldots}{1-\frac{1}{n}+\ldots}\frac{1}{x^2}$$

$$\therefore \quad \underset{n\to\infty}{\text{lt.}} \frac{u_n}{u_{n+1}} = \frac{1}{x^2}$$

Hence by ratio test the given series is

convergent if $\frac{1}{x^2} > 1$ i.e. $x^2 < 1$

divergent if $\frac{1}{x^2} < 1$ i.e. $x^2 > 1$

and the test fails if $\frac{1}{x^2} = 1$ i.e. $x^2 = 1$

Now when $x^2 = 1$,

$$u_n = \sqrt{n^2+1}-n$$

$$= n\left[\left(1+\frac{1}{n^2}\right)^{1/2}-1\right]$$

$$= n\left[1+\frac{1}{2}\frac{1}{n^2}-\frac{1}{8n^4}+\ldots-1\right]$$

$$= \frac{1}{2n} - \frac{1}{8n^3} + \ldots.$$

$$v_n = \frac{1}{n}$$

where v_n is n^{th} term of the auxiliary series Σv_n.

$$\frac{u_n}{v_n} = \frac{1}{2} - \frac{1}{8n^2} + \ldots$$

$$\therefore \lim_{n\to\infty} \frac{u_n}{v_n} = \frac{1}{x}$$

which is a fixed finite non-zero quantity.
Hence comparison test can be applied.

But $\Sigma \frac{1}{n}$ is divergent as $p = 1$ by p-test

Hence the given series is also divergent by comparison test.

Ex. 15. *Test the series*

$$\frac{x}{1.2} + \frac{x^2}{2.3} + \frac{x^3}{3.5} + \frac{x^4}{4.5} + \ldots$$

Sol. Here, $$u_n = \frac{x^n}{n(n+1)}$$

$$u_{n+1} = \frac{x^{n+1}}{(n+1)(n+2)}$$

$$\frac{u_n}{u_{n+1}} = \frac{n+2}{n} \frac{x^n}{x^{n+1}}$$

$$= \left(1 + \frac{2}{n}\right)\frac{1}{x}$$

$$\therefore \lim_{n\to\infty} \frac{u_n}{u_{n+1}} = \frac{1}{x}$$

Hence by ratio test the given series is

convergent if $\frac{1}{x} > 1$ i.e. $x < 1$

divergent if $\frac{1}{x} < 1$ i.e. $x > 1$

and the test fails if $\frac{1}{x} = 1$ i.e. $x = 1$

When $\quad x = 1,$

$$u_n = \frac{1}{n(n+1)}$$

$$v_n = \frac{1}{n^2}$$

where v_n is n^{th} term of the auxiliary series Σv_n.

$$\frac{u_n}{v_n} = \frac{n}{n+1} = \frac{1}{1+\frac{1}{n}}$$

$$\therefore \quad \underset{n\to\infty}{\text{lt.}} \frac{u_n}{v_n} = 1$$

which is a fixed finite non-zero quantity.

Hence comparison test can be applied.

But $\Sigma \frac{1}{n^2}$ is convergent as $p(= 2) > 1$ by p-test.

Hence the given series is also convergent by comparison test.

Ex. 16. *Test the series*

$$\frac{1}{2\sqrt{1}} + \frac{x^2}{3\sqrt{2}} + \frac{x^4}{4\sqrt{3}} + \frac{x^6}{5\sqrt{4}} + ...$$

Sol. Here, $\quad u_n = \frac{x^{2n-2}}{(n+1)\sqrt{n}}$

$$u_{n+1} = \frac{x^{2n}}{(n+2)\sqrt{n+1}}$$

$$\frac{u_n}{u_{n+1}} = \frac{n+2}{n+1}\frac{\sqrt{n+1}}{\sqrt{n}}\frac{x^{2n-2}}{x^{2n}}$$

$$= \frac{1+\frac{2}{n}}{1+\frac{1}{n}}\sqrt{1+\frac{1}{n}}\,\frac{1}{x^2}$$

$$\therefore \quad \underset{x\to\infty}{\text{lt.}} \frac{u_n}{u_{n+1}} = \frac{1}{x^2}$$

Hence by ratio test, the given series is

convergent if $\frac{1}{x^2} > 1$ i.e. $x^2 < 1$

divergent if $\frac{1}{x^2} < 1$ i.e. $x^2 > 1$

and the test fails if $\frac{1}{x^2} = 1$ i.e. $x^2 = 1$.

When $\quad x^2 = 1,$

$$u_n = \frac{1}{(n+1)\sqrt{n}}$$

$$v_n = \frac{1}{n\sqrt{n}} = \frac{1}{n^{3/2}}$$

where v_n is n^{th} term of the auxiliary series Σv_n.

$$\frac{u_n}{v_n} = \frac{1}{1+\frac{1}{n}}$$

$$\therefore \quad \underset{n\to\infty}{\text{lt.}} \frac{u_n}{v_n} = 1$$

which is a fixed finite non-zero quantity.

Hence comparison test can be applied.

But $\Sigma \frac{1}{n^{3/2}}$ is convergent as $p\ (= 3/2) > 1$ by p-test.

Hence the given series is also convergent by comparison test.

Ex. 17. *Test the series*

$$\Sigma \left(\frac{n^3+a}{2^n+a}\right)$$

Sol. Here, $\quad u_n = \frac{n^3+a}{2^n+a}$

$$u_{n+1} = \frac{(n+1)^3 + a}{2^{n+1} + a}$$

$$\frac{u_n}{u_{n+1}} = \frac{n^3 + a}{(n+1)^3 + a} \cdot \frac{2^{n+1} + a}{2^n + a}$$

$$= \frac{1 + \frac{a}{n^3}}{\left(1 + \frac{1}{n}\right)^3 + \frac{a}{n^3}} \cdot \frac{2 + \frac{a}{2^n}}{1 + \frac{a}{2^n}}$$

$$\therefore \lim_{n \to \infty} \frac{u_n}{u_{n+1}} = 2 > 1$$

Hence by ratio test, the given series Σu_n is convergent.

Ex. 18. *Examine the convergence of the series*

$$\frac{1}{x-1} + \frac{x}{x^2-1} + \frac{x^2}{x^3-1} + \ldots (x > 1)$$

Sol. The given series is

$$> \frac{1}{x} + \frac{x}{x^2} + \frac{x^2}{x^3} + \ldots \qquad (\because x > 1)$$

$$> \frac{1}{x} + \frac{1}{x} + \frac{1}{x} + \ldots$$

Let S_n be the sum of the n term of the series then

$$S_n > \frac{n}{x}$$

$$\therefore \lim_{n \to \infty} S_n \to \infty$$

Hence the given series is divergent.

Ex. 19. *Show that the series*

$$1 + \frac{\alpha+1}{\beta+1} + \frac{(\alpha+1)(2\alpha+1)}{(\beta+1)(2\beta+1)} + \frac{(\alpha+1)(2\alpha+1)(3\alpha+1)}{(\beta+1)(2\beta+1)(3\beta+1)} + \ldots$$

converges if $\beta > \alpha > 0$ and diverges

if $\alpha \geq \beta > 0$ $\qquad [\alpha > 0, \beta > 0]$

Sol. Here, $$u_n = \frac{(\alpha+1)(2\alpha+1)(3\alpha+1)...[(n-1)\alpha+1]}{(\beta+1)(2\beta+1)(3\beta+1)...[(n-1)\beta+1]}$$

$$\therefore \quad u_{n+1} = \frac{(\alpha+1)(2\alpha+1)...\{(n-1)\alpha+1\}(n\alpha+1)}{(\beta+1)(2\beta+1)...\{(n-1)\beta+1\}(n\beta+1)}$$

$$\frac{u_n}{u_{n+1}} = \frac{n\beta+1}{n\alpha+1}$$

$$= \frac{\beta + \frac{1}{n}}{\alpha + \frac{1}{n}}$$

$$\therefore \quad \underset{n\to\infty}{\text{lt.}} \frac{u_n}{u_{n+1}} = \frac{\beta}{\alpha}$$

Hence by ratio test, the given series is convergent if

$$\frac{\beta}{\alpha} > 1 \text{ i.e. } \beta > \alpha > 0 \qquad (\text{as } \alpha > 0)$$

and is divergent if $\frac{\beta}{\alpha} < 1$ i.e.

$$\beta < \alpha$$

or $$\alpha > \beta > 0 \qquad (\text{as } \beta > 0)$$

and test fails if $$\frac{\beta}{\alpha} = 1$$

i.e. $$\alpha = \beta > 0 \qquad (as\, \beta > 0)$$

Now when $\alpha = \beta$,

the given series becomes,

$$1 + 1 + 1 + ...$$

Let S_n be the sum of n terms of the series. Then

$$S_n = n$$

$$\therefore \quad \underset{n\to\infty}{\text{lt.}} S_n = \infty$$

Hence the given series is divergent when

$$\alpha = \beta > 0$$

as $$\alpha > 0 \text{ and } \beta > 0$$

Ex. 20. *Test the series*

$$\sum \frac{1}{x^n + x^{-u}}$$

Sol. Here,

$$u_n = \frac{1}{x^n + x^{-u}}$$

$$= \frac{x^n}{x^{2n} + 1}$$

$$u_{n+1} = \frac{x^{n+1}}{x^{2n+2} + 1}$$

$$\frac{u_n}{u_{n+1}} = \frac{x^{2n+2} + 1}{x^{2n} + 1} \frac{1}{x}$$

Let $x > 1$, then,

$$\frac{u_n}{u_{n+1}} = \frac{x^{2n+2}\left(1 + \frac{1}{x^{2n+2}}\right)}{x^{2n}\left(1 + \frac{1}{x^{2n}}\right)} \frac{1}{x}$$

$$= x\left[\frac{1 + \frac{1}{x^{2n+2}}}{1 + \frac{1}{x^{2n}}}\right]$$

$\because \lim_{n\to\infty} \frac{1}{x^{2n}} \to 0$ as $x > 1$

and $\lim_{n\to\infty}$ as $x > 1$

$$\therefore \lim_{n\to\infty} \frac{u_n}{u_{n+1}} = x$$

$$> 1$$

Hence by ratio test the given series is convergent.

Again, let $x < 1$, then

$$\lim_{n\to\infty} \frac{u_n}{u_{n+1}} = \lim_{n\to\infty} \frac{x^{2n+2} + 1}{x^{2n} + 1} \frac{1}{x}$$

$$= \frac{1}{x}$$

$$> 1 \qquad \text{(as } x<1)$$

$$\left| \begin{array}{ll} \because & x^{2n} \to \infty \\ \text{and} & x^{2n+2} \to \infty \end{array} \right\} \begin{array}{l} \text{as } n \to \infty \\ \because\ x < 1 \end{array}$$

Hence by ratio test the given series is convergent.

Now when $x = 1$, then the given series becomes

$$\frac{1}{2}+\frac{1}{2}+\frac{1}{2}+...$$

Let S_n be the sum of n terms of the series then

$$S_n = \frac{n}{2}$$

$$\therefore \quad \underset{n\to\infty}{\text{lt.}} S_n = \infty$$

Hence the series is divergent. Thus the given series is convergent.

when $x > 1$ or < 1 and divergent.

when $x = 1$.

Ex. 21. *Test the series*

$$\Sigma\,(3n - 1)\, 2^n.$$

Sol. Here, $\quad u_n = (3n - 1)\, 2^n$

$$u_{n+1} = (3n + 2)\, 2^{n+1}$$

$$\frac{u_n}{u_{n+1}} = \frac{3n-1}{3n+2}\,\frac{1}{2}$$

$$= \frac{3-\frac{1}{n}}{3+\frac{2}{n}}\,\frac{1}{2}$$

$$\therefore \quad \underset{n\to\infty}{\text{lt.}} \frac{u_n}{u_{n+1}} = \frac{1}{2} < 1$$

Hence the given series is divergent by ratio test.

Ex. 22. *Test the series*

$$\Sigma \frac{n\,!}{n^n}$$

Sol. Here, $u_n = \dfrac{n\,!}{n^n}$

$$u_{n+1} = \frac{n+1\,!}{(n+1)^{n+1}}$$

$$\frac{u_n}{u_{n+1}} = \frac{(n+1)^{n+1}}{n^n} \cdot \frac{n!}{n+1!}$$

$$= \frac{(n+1)^{n+1}}{n^n} \frac{1}{n+1}$$

$$= \frac{(n+1)^n}{n^n}$$

$$= \left(1 + \frac{1}{n}\right)^n$$

$$\therefore \quad \underset{n\to\infty}{\text{lt.}} \frac{u_n}{u_{n+1}} = e > 1 \qquad | \text{ as } e = 2.718$$

Hence by ratio test, the given series is convergent.

Ex. 23. *Test the series*

$$\Sigma \frac{x^n}{a+\sqrt{n}}$$

Sol. Here, $u_n = \dfrac{x^n}{a+\sqrt{n}}$

$$u_{n+1} = \frac{x^{n+1}}{a+\sqrt{n+1}}$$

$$\frac{u_n}{u_{n+1}} = \frac{a+\sqrt{n+1}}{a+\sqrt{n}} \frac{x^n}{x^{n+1}}$$

$$= \frac{a+\sqrt{n+1}}{a+\sqrt{n}} \frac{1}{x}$$

$$= \frac{\frac{a}{\sqrt{n}} + \sqrt{1+\frac{1}{n}}}{\frac{a}{\sqrt{n}}+1} \cdot \frac{1}{x}$$

$$\therefore \quad \underset{n\to\infty}{\text{lt.}} \frac{u_n}{u_{n+1}} = \frac{1}{x}$$

Hence by ratio test the given series is

convergent if $\frac{1}{x} > 1$ i.e. $x < 1$

divergent if $\frac{1}{x} < 1$ i.e. $x > 1$

and the test fails when $\frac{1}{x} = 1$ i.e. $x = 1$

Now when $x = 1$,

$$u_n = \frac{1}{a+\sqrt{n}}$$

$$v_n = \frac{1}{\sqrt{n}}$$

where v_n is n^{th} term of the auxiliary series Σv_n.

$$\frac{u_n}{v_n} = \frac{\sqrt{n}}{a+\sqrt{n}}$$

$$= \frac{1}{1+\frac{a}{\sqrt{n}}}$$

$$\therefore \quad \underset{n\to\infty}{\text{lt.}} \frac{u_n}{v_n} = 1$$

which is a fixed finite non-zero quantity.

Hence comparison test can be applied.

But $\Sigma \frac{1}{\sqrt{n}}$ is divergent as $p\left(=\frac{1}{2}\right) < 1$ by p-test.

Hence given series is also divergent,

when $x = 1$ by comparison test.

Ex. 24. *Test the series*

$$\frac{1}{1.2.3}+\frac{x}{4.5.6}+\frac{x^2}{7.8.9}+\ldots$$

Sol. Here,

$$u_n=\frac{x^{n-1}}{(3n-2)(3n-1)3n}$$

$$u_{n+1}=\frac{x^n}{(3n+1)(3n+2)(3n+3)}$$

$$\frac{u_n}{u_{n+1}}=\frac{(3n+1)(3n+2)(3n+3)}{(3n-2)(3n-1)3n}\frac{x^{n-1}}{x^n}$$

$$=\frac{\left(3+\frac{1}{n}\right)\left(3+\frac{2}{n}\right)\left(1+\frac{1}{n}\right)}{\left(3-\frac{2}{n}\right)\left(3-\frac{1}{n}\right)}\frac{1}{x}$$

$$\therefore \quad \underset{n\to\infty}{\text{lt.}} \quad \frac{u_n}{u_{n+1}}=\frac{1}{x}$$

Hence by ratio test, the given series is

convergent if $\frac{1}{x}>1$ i.e. $x<1$

divergent if $\frac{1}{x}<1$ i.e. $x>1$

and the test fails if $\frac{1}{x}=1$ i.e. $x=1$

Now when $x=1$

$$u_n=\frac{1}{(3n-2)(3n-1)3n}$$

$$v_n=\frac{1}{n^3}$$

where v_n is n^{th} term of the auxiliary series Σv_n.

$$\frac{u_n}{v_n}=\frac{n^3}{(3n-2)(3n-1)3n}$$

$$= \frac{n^2}{3(3n-2)(3n-1)}$$

$$= \frac{1}{3\left(3-\frac{2}{n}\right)\left(3-\frac{1}{n}\right)}$$

$$\therefore \quad \underset{n\to\infty}{\text{lt.}} \quad \frac{u_n}{v_n} = \frac{1}{27}$$

which is a fixed finite non-zero quantity.

Hence comparison test can be applied.

But $\Sigma \frac{1}{n^3}$ is convergent as $p(=3) > 1$ by p-test.

Hence the given series is also convergent,

when $x = 1$ by comparison test.

Ex. 25. *Show that the series*

$$1 + \frac{x \log a}{1!} + \frac{x^2 (\log a)^2}{2!} + \ldots$$

is convergent for all values of x.

Sol. Let the given series be denoted by Σu_n. Then

$$u_n = \frac{x^{n-1} (\log a)^{n-1}}{n-1!}$$

$$\therefore \quad u_{n+1} = \frac{x^n (\log a)^n}{n!}$$

$$\therefore \quad \frac{u_n}{u_{n+1}} = \frac{x^{n-1} (\log a)^{n-1}}{(n-1)!} \cdot \frac{n!}{x^n (\log a)^n}$$

$$= \frac{n}{x \log a}.$$

$$\therefore \quad \underset{n\to\infty}{\text{lt.}} \quad \frac{u_n}{u_{n+1}} = \underset{n\to\infty}{\text{lt.}} \quad \frac{n}{x \log a}$$

$$= \infty > 1 \text{ for all values of } x.$$

Hence by ratio test, the given series is convergent.

1.19 An Important Comparison Test

Statement. If Σu_n and Σv_n be two series of positive terms and if from and after some fixed terms,

(i) $\dfrac{u_n+1}{u_n} < \dfrac{v_n+1}{v_n}$ for all values of n, then Σu_n will be convergent if Σv_n is convergent, and if,

(ii) $\dfrac{u_n+1}{u_n} > \dfrac{v_n+1}{v_n}$ for all values of n, then Σu_n will be divergent if Σv_n is divergent.

Proof. Let the given series of +ve terms be

$$\Sigma u_n = u_1 + u_2 + u_3 + \ldots$$

and $$\Sigma v_n = v_1 + v_2 + v_3 + \ldots$$

Case I. We are given that,

$$\frac{u_n+1}{u_n} < \frac{v_n+1}{v_n}$$

for all values of n

$$\therefore \quad \frac{u_2}{u_1} < \frac{v_2}{v_1}$$

$$\frac{u_3}{u_2} < \frac{v_3}{v_2}$$

$$\frac{u_4}{u_3} < \frac{v_4}{v_3}$$

and so on.

Now, $$u_1 + u_2 + u_3 + \ldots$$

$$= u_n\left[1 + \frac{u_2}{u_1} + \frac{u_3}{u_1} + \ldots\right]$$

$$= u_1\left[1 + \frac{u_2}{u_1} + \frac{u_3}{u_2}.\frac{u_2}{u_1} + \ldots\right]$$

$$< u_1\left[1 + \frac{v_2}{v_1} + \frac{v_3}{v_2}.\frac{v_2}{v_1} + \ldots\right]$$

$$< \frac{u_1}{v_1}[v_1 + v_2 + v_3 + ...]$$

Hence Σu_n is convergent if Σv_n is convergent.

Case II. We are given that

$$\frac{u_{n+1}}{u_n} > \frac{v_{n+1}}{v_n}$$

for all values of n

$$\therefore \quad \frac{u_2}{u_1} > \frac{v_2}{v_1},$$

$$\frac{u_3}{u_2} > \frac{v_3}{v_2},$$

$$\frac{u_4}{u_3} > \frac{v_4}{v_3},$$

and so on.

Now, $u_1 + u_2 + u_3 + ...$

$$\therefore \quad = u_1\left[1 + \frac{u_2}{u_1} + \frac{u_3}{u_1} + ...\right]$$

$$= u_1\left[1 + \frac{u_2}{u_1} + \frac{u_3}{u_2}.\frac{u_2}{u_1}...\right]$$

$$> u_1\left[1 + \frac{v_2}{v_1} + \frac{v_3}{v_2}.\frac{v_2}{v_1}...\right]$$

$$> \frac{u_1}{v_1}\left[v_1 + v_2 + v_3 + ...\right]$$

Hence Σu_n is divergent if Σv_n is divergent.

1.20 The results obtained in 1.19 are very much used in many further articles (tests). For the sake of convenience, they are used in the following form :

If
$$\frac{u_n}{u_n + 1} > \frac{v_n}{v_n + 1}$$

and Σv_n is convergent then Σu_n is also convergent and if $$\frac{u_n}{u_n+1} > \frac{v_n}{v_n+1}$$

and Σv_n is divergent then Σu_n is also divergent.

1.21 Raabe's Test

Statement. The series Σu_n of positive terms is convergent or divergent according as

$$\underset{n\to\infty}{\text{lt.}}\ n\left(\frac{u_n}{u_{n+1}}-1\right) > 1 \text{ or } < 1$$

Proof.

Let the given series of positive terms be

$$\Sigma u_n = u_1 + u_2 + u_3 + \ldots$$

Compare this series with the auxiliary series

$$\Sigma v_n = \Sigma \frac{1}{n^p}$$

which by p-test is convergent or divergent according as

$$p > 1 \text{ or } p < 1$$

So, $$v_n = \frac{1}{n^p}$$

$$\therefore \quad v_{n+1} = \frac{1}{(n+1)^p}$$

$$\frac{v_n}{v_{n+1}} = \left(\frac{n+1}{n}\right)^p$$

$$= \left(1 + \frac{1}{n}\right)^p$$

$$= 1 + \frac{p}{n} + \frac{p(p-1)}{2\,!}\frac{1}{n^2} + \ldots.$$

(By Binomial Theorem)

$$\because \quad n\left(\frac{v_n}{v_{n+1}}-1\right) = p + \frac{p(p-1)}{2\,!}\frac{1}{n} + \ldots.$$

$$\therefore \quad \underset{n\to\infty}{\text{lt.}}\ n\left(\frac{v_n}{v_{n+1}}-1\right) = p \qquad \ldots..(1)$$

Now let $\frac{u_n}{u_{n+1}} > \frac{v_n}{v_{n+1}}$, then from (1)

$$\underset{n\to\infty}{\text{lt.}} \quad n\left(\frac{u_n}{u_{n+1}} - 1\right) > p$$

Again if Σv_n is convergent then $p > 1$,

Hence Σu_n is convergent if

$$\underset{n\to\infty}{\text{lt.}} \quad n\left(\frac{u_n}{u_{n+1}} - 1\right) > 1$$

Further, let $\frac{u_n}{u_{n+1}} < \frac{v_n}{v_{n+1}}$, then from (1)

$$\underset{n\to\infty}{\text{lt.}} \quad n\left(\frac{u_n}{u_{n+1}} - 1\right) < p$$

Again if Σv_n is divergent then $p < 1$,

Hence Σu_n is divergent if

$$\underset{n\to\infty}{\text{lt.}} \quad n\left(\frac{u_n}{u_{n+1}} - 1\right) < 1$$

Note 1 : Raabe's test is applied only when D' Alembert's ratio test fails, i.e. when

$$\underset{n\to\infty}{\text{lt.}} \quad \frac{u_n}{u_{n+1}} = 1$$

Note 2 : Raabe's test fails if

$$\underset{n\to\infty}{\text{lt.}} \quad n\left(\frac{u_n}{u_n + 1} - 1\right) = 1$$

1.22 Logarithmic Test

Statement. The series Σu_n of positive terms is convergent or divergent according as

$$\underset{n\to\infty}{\text{lt.}} \quad n \log \frac{u_n}{u_{n+1}} > 1 \text{ or } < 1$$

Proof.

Let the given series of positive terms be

$$\Sigma u_n = u_1 + u_2 + u_3 + \ldots$$

Compare this series with the auxiliary series

$$\Sigma v_n = \Sigma \frac{1}{n^p}$$

which by *p*-test is convergent or divergent according as

$$p > 1 \text{ or } p < 1$$

So,
$$v_n = \frac{1}{n^p}$$

$$v_{n+1} = \frac{1}{(n+1)^p}$$

$$\frac{v_n}{v_{n+1}} = \left(\frac{n+1}{n}\right)^p = \left(1 + \frac{1}{n}\right)^p$$

$$\log \frac{v_n}{v_{n+1}} = \log\left(1 + \frac{1}{n}\right)^p$$

$$= p \log\left(1 + \frac{1}{n}\right)$$

$$= p\left[\frac{1}{n} - \frac{1}{2n^2} + \frac{1}{3n^3} - \ldots\right]$$

or,
$$n \log \frac{v_n}{v_{n+1}} = p\left[1 - \frac{1}{2n} + \frac{1}{3n^2} + \ldots\right]$$

$$\therefore \quad \operatorname*{lt.}_{n\to\infty} n \log \frac{v_n}{v_{n+1}} = p \qquad \ldots\ldots(1)$$

Now let $\frac{u_n}{u_{n+1}} > \frac{v_n}{v_{n+1}}$ then from (1)

$$\operatorname*{lt.}_{n\to\infty} n \log \frac{u_n}{u_{n+1}} > p$$

Again, if Σv_n is convergent then $p > 1$,

Hence Σu_n is convergent if

$$\lim_{n\to\infty} n \log \frac{u_n}{u_{n+1}} > 1$$

Further, let $\dfrac{u_n}{u_{n+1}} < \dfrac{v_n}{v_{n+1}}$, then from (1)

$$\lim_{n\to\infty} n \log \frac{u_n}{u_{n+1}} < p$$

Again if Σv_n is divergent then $p < 1$,

Hence Σu_n is divergent if

$$\lim_{n\to\infty} n \log \frac{u_n}{u_{n+1}} < 1$$

Note 1 : Logarithmic test is applied when D'Alembert's ratio test fails and is preferred when e or powers involved.

Note 2 : Logarithmic test fails when

$$\lim_{n\to\infty} n \log \frac{u_n}{u_{n+1}} = 1.$$

1.23 Cauchy's Condensation Test

Statement. If the function $f(n)$ is positive for all positive integral values of n and continually decreases as n increases, and if m be a positive integer greater than 1, then the two infinite series

$$\Sigma f(n) = f(1) + f(2) + f(3) + \ldots + f(n) + \ldots$$

and $\Sigma m^n f(m^n) = mf(m) + m^2 f(m^2) + m^3 f(m^3) + \ldots + m^n f(m^n) + \ldots$ are either both convergent or divergent.

Proof.

Let the terms of the series $\Sigma f(n)$ be arranged as follows :

$$\begin{aligned}\Sigma f(n) = {} & f(1) + f(2) + \ldots + f(m) \\ & + f(m+1) + f(m+2) + \ldots + f(m^2) \\ & + f(m^2+1) + f(m^2+2) + \ldots + f(m^3) \\ & + \ldots\ldots\ldots\ldots\ldots\ldots \\ & \ldots\ldots\ldots\ldots\ldots\ldots\end{aligned}$$

$$+ f(m^k + 1) + f(m^k + 2) + \ldots + f(m^{k+1})$$
$$+ \ldots\ldots\ldots\ldots\ldots\ldots\ldots\ldots\ldots\ldots$$
$$\ldots\ldots\ldots\ldots\ldots\ldots\ldots\ldots\ldots\ldots$$

Since $f(n)$ decreases continually as n increases, therefore, we can write $\Sigma f(n)$ as,

$$\Sigma f(n) > mf(m) + (m^2 - m) f(m^2) + (m^3 - m^2) f(m^3) + \ldots + (m^{k+1} - m^k) f(m^{k+1}) + \ldots$$

or,
$$\Sigma f(n) > f(m) - f(m) + mf(m) + (m^2 - m) f(m^2) + (m^3 - m^2) f(m^3) + \ldots + (m^{k+1} - m^k) f(m^{k+1}) + \ldots$$

[adding and subtracting the term $f(m)$ in R.H.S.]

or,
$$\Sigma f(n) > f(m) + (m-1) f(m) + m(m-1) f(m^2) + m^2(m-1) f(m^3) + \ldots + m^k(m-1) f(m^{k+1}) + \ldots$$

$$> f(m) + (m-1)\,[\,f(m) + mf(m^2) + m^2 f(m^3) + \ldots + m^k f(m^{k+1}) + \ldots]$$

$$> f(m) + \frac{m-1}{m}\,[mf(m) + m^2 f(m^2) + m^3 f(m^3) + \ldots + m^{k+1} f(m^{k+1}) + \ldots]$$

$$> f(m) + \frac{m-1}{m} \Sigma m^n f(m^n)$$

Hence $\Sigma f(n)$ is divergent if $\Sigma m^n f(m^n)$ is divergent.

Again, we can write $\Sigma f(n)$ as,

$$\Sigma f(n) < mf(1) + (m^2 - m) f(m) + (m^3 - m^2) f(m^2) + \ldots + (m^{k+1} + m^k) f(m^k) + \ldots$$

$$< mf(1) + (m-1)[mf(m) + m^2 f(m^2) + \ldots + m^k f(m^k) + \ldots]$$

$$< mf(1) + (m-1)\, m^n f(m^n)$$

Hence $\Sigma f(n)$ is convergent if $\Sigma m^n f(m^n)$ is convergent.

1.24 An Important Auxilary Series

Statement. The series $\dfrac{1}{n(\log n)^p}$ is convergent or divergent according as $p > 1$ or $p \leq 1$ $(n \neq 1)$

Proof.

We shall apply Cauchy's condensation test to prove this result.

Here,

$$f(n) = \frac{1}{n(\log n)p}$$

clearly $f(n) \to 0$ as $n \to \infty$

Let m be a +ve integer > 1, then,

$$m^n f(m^n) = m^n \frac{1}{m^n (\log m^n)^p}$$

$$= \frac{1}{(n \log m)^p}$$

$$= \frac{1}{(\log m)^p} \cdot \frac{1}{n^p}$$

$$\therefore \quad \Sigma m^n f(m^n) = \frac{1}{(\log m)^p} \Sigma \frac{1}{n^p}$$

as $(\log m)^p$ is constant.

Now $\Sigma m^n f(m^n)$ will be convergent or divergent according as $\Sigma \dfrac{1}{n^p}$ is convergent or divergent i.e.

According as $p > 1$ or $p \leq 1$

and consequently $\Sigma \dfrac{1}{n(\log p)^p}$ will be convergent or divergent according as $p < 1$ or $p \leq 1$ by Cauchy's condensation test.

1.25 De Morgan and Bertrand's Test

Statement. The series Σu_n of positive terms is convergent or divergent according as

$$\underset{n \to \infty}{\text{lt.}} \log n \left[n \left(\frac{u_n}{u_{n+1}} - 1 \right) - 1 \right] > 1 \text{ or } < 1.$$

Proof.

Let the given series of positive terms be

$$\Sigma u_n = u_1 + u_2 + u_3 + \ldots$$

Compare this series with the auxiliary series

$$\Sigma v_n = \Sigma \frac{1}{n(\log n)^p}$$

which is convergent when $p > 1$ and divergent when $p \leq 1$. $(n \neq 1)$

so, $$v_n = \frac{1}{n(\log n)^p}$$

$$v_{n+1} = \frac{1}{(n+1)\,[\log (n+1)]^p}$$

$$\frac{n}{v_{n+1}} = \frac{n+1}{n}\left\{\frac{\log (n+1)}{\log n}\right\}^p$$

$$= \left(1+\frac{1}{n}\right)\left[\frac{\log\left\{n\left(1+\frac{1}{n}\right)\right\}}{\log n}\right]^p$$

$$= \left(1+\frac{1}{n}\right)\left[\frac{\log n + \log\left(1+\frac{1}{n}\right)}{\log n}\right]^p$$

$$= \left(1+\frac{1}{n}\right)\left[1+\frac{\log\left(1+\frac{1}{n}\right)}{\log n}\right]^p$$

$$= \left(1+\frac{1}{n}\right)\left[1+\frac{1}{\log n}\left\{\frac{1}{n}-\frac{1}{2n^2}+\ldots\right\}\right]^p$$

$$= \left(1+\frac{1}{n}\right)\left(1+\frac{1}{n\log n}-\frac{1}{2n^2\log n}+\ldots\right)^p$$

$$= \left(1 + \frac{1}{n}\right)\left(1 + \frac{p}{n \log n} + \ldots\right)$$

(By Binomial Expansion)

$$= 1 + \frac{1}{n} + \frac{p}{n \log n} + \ldots$$

$$\therefore \quad \log n \left[n\left(\frac{v_n}{v_{n+1}} - 1\right) - 1 \right]$$

$$= p + \text{terms containing powers of } \frac{1}{n}$$

$$\therefore \quad \underset{n \to \infty}{\text{lt.}} \quad \log n \left[n\left(\frac{v_n}{v_{n+1}} - 1\right) - 1 \right] = p \qquad \ldots.(1)$$

Now let $\frac{u_n}{u_{n+1}} > \frac{v_n}{v_{n+1}}$, then from (1)

$$\therefore \quad \underset{n \to \infty}{\text{lt.}} \quad \left[n\left(\frac{u_n}{u_{n+1}} - 1\right) - 1 \right] \log n > p$$

Again, if Σv_n is convergent then $p > 1$, hence Σu_n is convergent if

$$\therefore \quad \underset{n \to \infty}{\text{lt.}} \quad \log n \left[n\left(\frac{u_n}{u_{n+1}} - 1\right) - 1 \right] > 1$$

Further, let

$$\frac{u_n}{u_{n+1}} < \frac{v_n}{v_{n+1}} \text{ then from (1)}$$

$$\underset{n \to \infty}{\text{lt.}} \quad \left[n\left(\frac{u_n}{u_{n+1}} - 1\right) - 1 \right] \log n < p$$

Again if Σv_n is divergent then $p < 1$, hence Σu_n is divergent if

$$\underset{n \to \infty}{\text{lt.}} \quad \left[n\left(\frac{u_n}{u_{n+1}} - 1\right) - 1 \right] \log n < 1$$

Note 1 : This test is applied only when Raabe's Test fails, i.e.

$$\text{when} \quad \underset{n\to\infty}{\text{lt.}} \; n\left(\frac{u_n}{u_{n+1}}-1\right)=1$$

Note 2 : This test fails if

$$\underset{n\to\infty}{\text{lt.}} \; \log n\left[n\left(\frac{u_n}{u_{n+1}}-1\right)-1\right]=1$$

1.26. Special Logarithmic Test

Statement. The series Σu_n of positive terms is convergent or divergent according as

$$\underset{n\to\infty}{\text{lt.}} \left[\left\{n\log\frac{u_n}{u_{n+1}}-1\right\}\log n\right] >1 \text{ or } <1$$

Proof.

Let the given series of positive terms be $\Sigma u_n = u_1 + u_2 + u_3 + \ldots$
Compare this series with the auxiliary series

$$\Sigma v_n = \Sigma \frac{1}{n(\log n)^p}$$

which is convergent when $p > 1$ and divergent when $p < 1$ $(n \neq 1)$

So, $$v_n = \frac{1}{n(\log n)^p}$$

$$v_{n+1} = \frac{1}{(n+1)\{\log(n+1)\}^p}$$

$$\therefore \quad \frac{v_n}{v_{n+1}} = 1+\frac{1}{n}+\frac{p}{n\log n}+\ldots \quad \text{(as in 1.25}$$

$$\therefore \quad \log\frac{v_n}{v_{n+1}} = \log\left(1+\frac{1}{n}+\frac{1}{n\log n}+\ldots\right)$$

$$=\frac{1}{n}+\frac{p}{n\log n}+\ldots$$

$$\therefore \quad \left[n \log \frac{v_n}{v_{n+1}} - 1 \right] \log n$$

$$= p + \text{terms containing powers of } \frac{1}{n}$$

$$\underset{n \to \infty}{\text{lt.}} \; n \left[\log \frac{v_n}{v_{n+1}} - 1 \right] \log n = p \qquad \text{...(1)}$$

Now let, $\frac{u_n}{u_{n+1}} > \frac{v_n}{v_{n+1}}$, then from (1)

$$\underset{n \to \infty}{\text{lt.}} \left[n \log \frac{u_n}{u_{n+1}} - 1 \right] \log n > p$$

Again if Σv_n is convergent, then $p > 1$,

Hence Σu_n is convergent if

$$\underset{n \to \infty}{\text{lt.}} \left[n \log \frac{u_n}{u_{n+1}} - 1 \right] \log n > 1$$

Further let $\frac{u_n}{u_{n+1}} < \frac{v_n}{v_{n+1}}$, then from (1)

$$\underset{n \to \infty}{\text{lt.}} \left[n \log \frac{u_n}{u_{n+1}} - 1 \right] \log n < p$$

Again if Σv_n is divergent then $p < 1$,

Hence Σu_n is divergent if

$$\underset{n \to \infty}{\text{lt.}} \left[n \log \frac{u_n}{u_{n+1}} - 1 \right] \log n < 1$$

Note 1 : This test is applied only when logarithmic test fails i.e.

$$\text{when} \quad \underset{n \to \infty}{\text{lt.}} \; n \log \frac{u_n}{u_{n+1}} = 1$$

Note 2 : This test fails when

$$\underset{n \to \infty}{\text{lt.}} \left[n \log \frac{u_n}{u_{n+1}} - 1 \right] \log n = 1.$$

Illustrative Examples

(Based on Raabe's Test)

Ex. 1. *Test the convergence of the series*

$$\frac{1^2}{4^2}+\frac{1\cdot 5^2}{4^2\,8^2}+\frac{1^2\cdot 5^2\cdot 9^2}{4^2\cdot 8^2\cdot 12^2}+\frac{1^2\cdot 5^2\,9^2\,13^2}{4^2\cdot 8^2\cdot 12^2\cdot 16^2}+....$$

Sol. Here $$u_n=\frac{1^2\cdot 5^2\cdot 9^2\cdot 13^2 ...(4n-3)^2}{4^2\cdot 8^2\cdot 12^2\cdot 16^2 ...(4n)^2}$$

$$\therefore \quad u_{n+1}=\frac{1^2\cdot 5^2\cdot 9^2\cdot 13^2 ...(4n-3)^2\,(4n+1)^2}{4^2\cdot 8^2\cdot 12^2\cdot 16^2 ...(4n)^2\,(4n+4)^2}$$

$$\therefore \quad \frac{u_n}{u_{n+1}}=\left(\frac{4n+4}{4n+1}\right)^2=\left[\frac{4+\frac{4}{n}}{4+\frac{1}{n}}\right]^2$$

$$\therefore \quad \underset{n\to\infty}{\text{lt.}}\ \frac{u_n}{u_{n+1}}=1$$

Hence ratio test fails.

Now applying Raabe's Test, we have

$$\frac{u_n}{u_{n+1}}=\left(\frac{4n+4}{4n+1}\right)^2=\frac{16n^2+32n+16}{16n^2+8n+1}$$

$$\Rightarrow \quad \frac{u_n}{u_{n+1}}-1=\frac{24n+15}{16n^2\,8n+1}$$

$$\Rightarrow \quad n\left(\frac{u_n}{u_n}-1\right)=\frac{24n^2+15n}{16n^2+8n+1}$$

$$=\frac{24+\frac{15}{n}}{16+\frac{8}{n}+\frac{1}{n^2}}$$

$$\therefore \quad \underset{n\to\infty}{\text{lt.}}\ n\ \frac{u_n}{u_{n+1}}-1=\frac{24}{16}=\frac{3}{2}>1$$

Hence the given series is convergent.

Ex. 2. *Test the series*

$$1+\frac{2}{3}\cdot\frac{1}{4}+\frac{2\cdot 4}{3\cdot 5}\cdot\frac{1}{6}+\frac{2\cdot 4\cdot 6}{3\cdot 5\cdot 7}\cdot\frac{1}{8}+\ldots$$

Sol. Leaving the first term, let the given series be denoted by Σu_n. Then,

$$u_n=\frac{2\cdot 4\cdot 6\ldots(2n)}{3\cdot 5\cdot 7\ldots(2n+1)}\,\frac{1}{2n+2}$$

$$\therefore\quad u_{n+1}=\frac{2\cdot 4\cdot 6\ldots\,2n\cdot(2n+2)}{3\cdot 5\cdot 7\ldots(2n+1)(2n+3)}\,\frac{1}{2n+4}$$

$$\frac{u_n}{u_{n+1}}=\frac{2n+3}{2n+2}\,\frac{2n+4}{2n+2}$$

$$=\frac{\left(2+\frac{3}{n}\right)\left(2+\frac{4}{n}\right)}{\left(2+\frac{2}{n}\right)\left(2+\frac{2}{n}\right)}$$

$$\therefore\quad \underset{n\to\infty}{\text{lt.}}\ \frac{u_n}{u_{n+1}}=1$$

Hence ratio test fails.

Now applying Raabe's test, we have

$$\frac{u_n}{u_{n+1}}=\frac{4n^2+14n+12}{4n^2+8n+4}$$

$$\frac{u_n}{u_{n+1}}-1=\frac{6n+8}{4n^2+8n+4}$$

$$n\left(\frac{u_n}{u_{n+1}}-1\right)=\frac{6n^2+8n}{4n^2+8n+4}=\frac{6+\frac{8}{n}}{4+\frac{8}{n}+\frac{4}{n^2}}$$

$$\therefore\quad \underset{n\to\infty}{\text{lt.}}\ n\left(\frac{u_n}{u_{n+1}}-1\right)=\frac{6}{4}=\frac{3}{2}>1$$

Hence the given series is convergent.

Ex. 3. *Test the series*

$$1+\frac{2^2}{3\cdot 4}+\frac{2^2\cdot 4^2}{3\cdot 4\cdot 5\cdot 6}+\frac{2^2\cdot 4^2\cdot 6^2}{3\cdot 4\cdot 5\cdot 6\cdot 7\cdot 8}+\ldots$$

Sol. Leaving the first term which will not affect the nature of the series. Let the given series be denoted by Σu_n. Then,

$$u_n=\frac{2^2\cdot 4^2\cdot 6^2\ldots(2n)^2}{3\cdot 4\cdot 5\cdot 6\cdot 7\cdot 8\ldots(2n+1)(2n+2)}$$

$$\therefore \quad u_{n+1}=\frac{2^2\cdot 4^2\cdot 6^2\ldots(2n)^2\,(2n+2)^2}{3\cdot 4\cdot 5\cdot 6\cdot 7\cdot 8\ldots(2n+1)(2n+2)(2n+3)(2n+4)}$$

$$\therefore \quad \frac{u_n}{u_{n+1}}=\frac{(2n+3)(2n+4)}{(2n+2)^2}$$

$$=\frac{\left(2+\frac{3}{n}\right)\left(2+\frac{4}{n}\right)}{\left(2+\frac{2}{n}\right)^2}$$

$$\therefore \quad \underset{n\to\infty}{\text{lt.}}\ \frac{u_n}{u_{n+1}}=1$$

Hence ratio test fails.

Now applying Raabe's test, we have

$$\frac{u_n}{u_{n+1}}=\frac{4n^2+14n+12}{4n^2+8n+4}$$

$$\frac{u_n}{u_{n+1}}-1=\frac{6n+8}{4n^2+8n+4}$$

$$n\left(\frac{u_n}{u_{n+1}}-1\right)=\frac{6n^2+8}{4n^2+8n+4}$$

$$=\frac{6+\frac{8}{n}}{4+\frac{8}{n}+\frac{4}{n^2}}$$

$$\therefore \quad \underset{n\to\infty}{\text{lt.}}\ n\left(\frac{u_n}{u_{n+1}}-1\right)=\frac{6}{4}=\frac{3}{2}>1$$

Hence the given series is convergent.

Ex. 4. *Test the series*

$$\frac{x}{1}+\frac{1}{2}\cdot\frac{x^3}{3}+\frac{1\cdot 3}{2\cdot 4}\frac{x^5}{5}+\frac{1\cdot 3\cdot 5}{2\cdot 4\cdot 6}\frac{x^7}{7}+...$$

Sol. Leaving the first term, let the given series be denoted by Σu_n. Then,

$$u_n=\frac{1\cdot 3\cdot 5...(2n-1)}{2\cdot 4\cdot 6...2n}\frac{x^{2n+1}}{2n+1}$$

$$u_{n+1}=\frac{1\cdot 3\cdot 5...(2n-1)(2n+1)}{2\cdot 4\cdot 6...(2n)(2n+2)}\frac{x^{2n+3}}{2n+3}$$

$$\frac{u_n}{u_{n+1}}=\frac{2n+2}{2n+1}\cdot\frac{2n+3}{2n+1}\cdot\frac{1}{x^2}$$

$$=\frac{4n^2+10n+6}{4n^2+4n+1}\cdot\frac{1}{x^2}$$

$$=\frac{4+\frac{10}{n}+\frac{6}{n^2}}{4+\frac{4}{n}+\frac{1}{n^2}}\cdot\frac{1}{x^2}$$

$$\therefore \quad \underset{n\to\infty}{\text{lt.}}\ \frac{u_n}{u_{n+1}}=\frac{1}{x^2}$$

Hence by ratio test the given series is

convergent if $\frac{1}{x^2}>1$ i.e. $x^2<1$

divergent if $\frac{1}{x^2}<1$ i.e. $x^2>1$

and the test fails if $\frac{1}{x^2}=1$ i.e. $x^2=1$

When $x^2=1$,

we have,

$$\frac{u_n}{u_{n+1}}=\frac{4n^2+10n+6}{4n^2+4n+1}$$

Let us apply Raabe's test,

$$\frac{u_n}{u_{n+1}}-1=\frac{6n+5}{4n^2+4n+1}$$

$$n\left(\frac{u_n}{u_{n+1}}-1\right)=\frac{6n^2+5n}{4n^2+4n+1}$$

$$=\frac{6+\frac{5}{n}}{4+\frac{4}{n}+\frac{1}{n^2}}$$

$$\therefore \quad \underset{n\to\infty}{\text{lt.}}\ n\left(\frac{u_n}{u_{n+1}}-1\right)=\frac{6}{4}=\frac{3}{2}>1$$

Hence the given series is convergent when $x^2 = 1$.

Ex. 5. *Test for convergence the series*

$$1+\frac{x}{2}+\frac{1\cdot 3}{2\cdot 4}\cdot x^2+\frac{1\cdot 3\cdot 5}{2\cdot 4\cdot 6}\cdot x^3+\ldots, x>0$$

Sol. Neglecting the first term, let the given series be denoted by Σu_n. Then,

$$u_n=\frac{1\cdot 3\cdot 5\ldots(2n-1)}{2\cdot 4\cdot 6\ldots 2n}\cdot x^n$$

$$\therefore \quad u_{n+1}=\frac{1\cdot 3\cdot 5\ldots(2n-1)(2n+1)}{2\cdot 4\cdot 6\ldots 2n(2n+2)}\cdot x^{n+1}$$

$$\therefore \quad \frac{u_n}{u_{n+1}}=\frac{2n+2}{2n+1}\cdot\frac{1}{x}=\frac{2+\frac{2}{n}}{2+\frac{1}{n}}\cdot\frac{1}{x}$$

$$\therefore \quad \underset{n\to\infty}{\text{lt.}}\ \frac{u_n}{u_{n+1}}=\frac{1}{x}$$

Hence by ratio test the given series is

convergent if $\frac{1}{x}>1$ i.e. $x<1$,

divergent if $\frac{1}{x}<1$ i.e. $x>1$,

and the test fails if $\frac{1}{x}=1$ i.e. $x=1$

When $x = 1$,

$$\frac{u_n}{u_{n+1}} = \frac{2n+2}{2n+1}$$

Applying Raabe's test, we have

$$\frac{u_n}{u_{n+1}} - 1 = \frac{2n+2}{2n+1} - 1 = \frac{1}{2n+1}$$

$$\therefore \quad n\left(\frac{u_n}{u_{n+1}} - 1\right) = \frac{n}{2n+1} = \frac{1}{2+\frac{1}{n}}$$

$$\therefore \quad \underset{n\to\infty}{\text{lt.}}\ n\left(\frac{u_n}{u_{n+1}} - 1\right) = \frac{1}{2} < 1$$

Hence the given series is divergent when $x = 1$.

Ex. 6. *Test the series*

$$1 + \frac{3}{7}x + \frac{3.6}{7.10}x^2 + \frac{3.6.9}{7.10.13}x^3 + \ldots, x > 0.$$

Sol. Leaving the first term, let the given series be denoted by Σu_n. Then,

$$u_n = \frac{3\cdot 6\cdot 9 \ldots 3n}{7\cdot 10\cdot 13 \ldots (3n+4)}x^n$$

$$u_{n+1} = \frac{3\cdot 6\cdot 9 \ldots 3n(3n+3)}{7\cdot 10\cdot 13 \ldots (3n+4)(3n+7)}x^{n+1}$$

$$\frac{u_n}{u_{n+1}} = \frac{3n+7}{3n+3}\frac{1}{x} = \frac{3+\frac{7}{n}}{3+\frac{3}{n}}\frac{1}{x}$$

$$\therefore \quad \underset{n\to\infty}{\text{lt.}}\ \frac{u_n}{u_{n+1}} = \frac{1}{x}$$

Hence by ratio test the given series is

convergent if $\frac{1}{x} > 1$ i.e. $x < 1$

divergent if $\frac{1}{x} < 1$ i.e. $x > 1$

and the test fails if $\frac{1}{x} = 1$ i.e. $x = 1$

When $x = 1$,

$$\frac{u_n}{u_{n+1}} = \frac{3n+7}{3n+3}$$

Now apply Raabe's test, we have

$$\frac{u_n}{u_{n+1}} - 1 = \frac{3n+7}{3n+3} - 1 = \frac{4}{3n+3}$$

$$\therefore \quad n\left(\frac{u_n}{u_{n+1}} - 1\right) = \frac{4n}{3n+3} = \frac{4}{3+\frac{3}{n}}$$

$$\therefore \quad \underset{n\to\infty}{\text{lt.}} \left(\frac{u_n}{u_{n+1}} - 1\right) = \frac{4}{3} > 1$$

Hence Σu_n is convergent when $x = 1$.

Ex. 7. *Test the series*

$$2 + \frac{3}{2}x + \frac{4}{3}x^2 + \frac{5}{4}x^3 + \ldots,\ x>0$$

Sol. Here, $$u_n = \frac{n+1}{n}x^{n-1}$$

$$\therefore \quad u_{n+1} = \frac{n+2}{n+1}x^n$$

$$\therefore \quad \frac{u_n}{u_{n+1}} = \frac{(n+1)^2}{n(n+2)}\frac{1}{x} = \frac{\left(1+\frac{1}{n}\right)^2}{1+\frac{2}{n}} \cdot \frac{1}{x}$$

$$\therefore \quad \underset{n\to\infty}{\text{lt.}} \frac{u_n}{u_{n+1}} = \frac{1}{x}$$

Hence by ratio test, the given series is

convergent if $\frac{1}{x} > 1$ i.e. $x < 1$,

divergent if $\frac{1}{x} < 1$ i.e. $x > 1$

and the test fails if $\frac{1}{x} = 1$ i.e. $x = 1$

When $x = 1$,

$$\frac{u_n}{u_{n+1}} = \frac{(n+1)^2}{n(n+2)} = \frac{n^2+2n+1}{n^2+2n}$$

Applying Raabe's test, we have

$$\frac{u_n}{u_{n+1}} - 1 = \frac{1}{n^2+2n}$$

$$\therefore \quad n\left(\frac{u_n}{u_{n+1}} - 1\right) = \frac{n}{n^2+2n} = \frac{1}{n+2}$$

$$\therefore \quad \underset{n\to\infty}{\text{lt.}}\ n\left(\frac{u_n}{u_{n+1}} - 1\right) = \frac{1}{\infty} = 0 < 1$$

Hence the given series is divergent when $x = 1$.

Ex. 8. *Test the series*

$$1 + \frac{1}{2}\frac{x^2}{4} + \frac{1\cdot 3\cdot 5}{2\cdot 4\cdot 6}\frac{x^4}{8} + \frac{1\cdot 3\cdot 5\cdot 7\cdot 9}{2\cdot 4\cdot 6\cdot 8\cdot 10}\frac{x^6}{12} + \ldots.$$

Sol. Neglecting the first term let the given series be denoted by Σu_n. Then,

$$u_n = \frac{1.3.5...(4n-3)}{2.4.6...(4n-2)}\frac{x^{2n}}{4n}$$

$$\therefore \quad u_{n+1} = \frac{1.3.5...(4n-3)(4n-1)(4n+1)}{2.4.6...(4n-2)(4n)(4n+2)}\frac{x^{2n+2}}{4n+4}$$

$$\therefore \quad \frac{u_n}{u_{n+1}} = \frac{(4n)(4n+2)}{(4n-1)(4n+1)}\frac{4n+4}{4n}\frac{1}{x^2}$$

$$= \frac{(4n+2)(4n+4)}{(4n-1)(4n+1)}\frac{1}{x^2}$$

$$= \frac{\left(4+\frac{2}{n}\right)\left(4+\frac{4}{n}\right)}{\left(4-\frac{1}{n}\right)\left(4+\frac{1}{n}\right)} \frac{1}{x^2}$$

$$\therefore \quad \underset{n\to\infty}{\text{lt.}} \quad \frac{u_n}{u_{n+1}} = \frac{1}{x^2}$$

Hence by ratio test, the given series is

convergent if $\frac{1}{x^2} > 1$ i.e. $x^2 < 1$

divergent if $\frac{1}{x^2} < 1$ i.e. $x^2 > 1$

and the test fails if $\frac{1}{x^2} = 1$ i.e. $x^2 = 1$.

when $x^2 = 1$,

$$\frac{u_n}{u_{n+1}} = \frac{(4n+2)(4n+4)}{(4n-1)(4n+1)}$$

$$= \frac{16n^2+24n+8}{16n^2-1}$$

Now apply Raabe's test, we have

$$\frac{u_n}{u_{n+1}} - 1 = \frac{24n+9}{16n^2-1}$$

$$\therefore \quad n\left(\frac{u_n}{u_{n+1}} - 1\right) = \frac{24n^2+9n}{16n^2-1} = \frac{24+\frac{9}{n}}{16-\frac{1}{n^2}}$$

$$\therefore \quad \underset{n\to\infty}{\text{lt.}} \quad n\left(\frac{u_n}{u_{n+1}} - 1\right) = \frac{24}{16} = \frac{3}{2} > 1$$

Hence Σu_n is convergent when $x^2 = 1$.

Ex. 9. *Test the series*

$$x^2+\frac{2^2}{3\cdot 4}x^4+\frac{2^2\cdot 4^2}{3\cdot 4\cdot 5\cdot 6}x^6+\frac{2^2\cdot 4^2\cdot 6^2}{3\cdot 4\cdot 5\cdot 6\cdot 7\cdot 8}x^8+\ldots$$

Sol. Neglecting the first term, let the given series be denoted by Σu_n. Then,

$$u_n=\frac{2^2\cdot 4^2\cdot 6^2 \ldots\ (2n)^2}{3\cdot 4\cdot 5\cdot 6 \ldots\ (2n+1)(2n+2)}\,x^{2n+2}$$

$$\therefore \quad u_{n+1}=\frac{2^2\cdot 4^2\cdot 6^2 \ldots (2n)^2\,(2n+2)^2}{3\cdot 4\cdot 5\cdot 6 \ldots (2n+1)(2n+2)(2n+3)(2n+4)}\,x^{2n+4}$$

$$\therefore \quad \frac{u_n}{u_{n+1}}=\frac{(2n+3)(2n+4)}{(2n+2)^2}\,\frac{1}{x^2}$$

$$=\frac{\left(2+\frac{3}{n}\right)\left(2+\frac{4}{n}\right)}{\left(2+\frac{2}{n}\right)^2}\,\frac{1}{x^2}$$

$$\therefore \quad \operatorname*{lt.}_{n\to\infty}\ \frac{u_n}{u_{n+1}}=\frac{1}{x^2}$$

Hence by ratio test, Σu_n is

convergent if $\frac{1}{x^2}>1$ i.e. $x^2<1$

divergent if $\frac{1}{x^2}<1$ i.e. $x^2>1$

and the test fails when $\frac{1}{x^2}=1$ i.e. $x^2=1$

when $\quad x^2=1$,

we have

$$\frac{u_n}{u_{n+1}}=\frac{(2n+3)(2n+4)}{(2n+2)^2}$$

$$=\frac{4n^2+14n+12}{4n^2+8n+4}$$

Now applying Raabe's test, we have

$$\therefore \quad \frac{u_n}{u_{n+1}} - 1 = \frac{6n+8}{4n^2+8n+4}$$

$$\therefore \quad n\left(\frac{u_n}{u_{n+1}} - 1\right) = \frac{6n^2+8n}{4n^2+8n+4}$$

$$= \frac{6+\frac{8}{n}}{4+\frac{8}{n}+\frac{4}{n^2}}$$

$$\therefore \quad \underset{n\to\infty}{\text{lt.}}\ n\left(\frac{u_n}{u_{n+1}} - 1\right) = \frac{6}{4} = \frac{3}{2} > 1$$

Hence Σu_n is convergent when $x^2 = 1$.

Ex. 10. *Test the series*

$$\Sigma u_n = \frac{1\cdot 3\cdot 5 \ldots (2n-1)}{2\cdot 4\cdot 6 \ldots 2n}\frac{x^{2n}}{2n}$$

Sol. Here,

$$u_n = \frac{1\cdot 3\cdot 5 \ldots (2n-1)}{2\cdot 4\cdot 6 \ldots 2n}\frac{x^{2n}}{2n}$$

$$\therefore \quad u_{n+1} = \frac{1\cdot 3\cdot 5 \ldots (2n-1)(2n+1)}{2\cdot 4\cdot 6 \ldots 2n(2n+2)}\frac{x^{2n+2}}{2n+2}$$

$$\frac{u_n}{u_{n+1}} = \frac{2n+2}{2n+1}\cdot\frac{2n+2}{2n}\cdot\frac{1}{x^2}$$

$$= \frac{\left(2+\frac{2}{n}\right)\left(2+\frac{2}{n}\right)}{\left(2+\frac{1}{n}\right)(2)}\frac{1}{x^2}$$

$$\therefore \quad \underset{n\to\infty}{\text{lt.}}\ \frac{u_n}{u_{n+1}} = \frac{1}{x^2}$$

Hence by ratio test Σu_n is

convergent if $\frac{1}{x^2} > 1$ i.e. $x^2 < 1$

divergent if $\frac{1}{x^2} < 1$ i.e. $x^2 > 1$

and the test fails if $\frac{1}{x^2} = 1$ i.e. $x^2 = 1$

when $x^2 = 1$,

$$\frac{u_n}{u_{n+1}} = \frac{(2n+2)(2n+2)}{(2n+1)\,2n} = \frac{4n^2+8n+4}{4n^2+2n}$$

$$\frac{u_n}{u_{n+1}} - 1 = \frac{6n+4}{4n^2+8n+4}$$

$$\therefore \quad n\left(\frac{u_n}{u_{n+1}} - 1\right) = \frac{6n^2+4n}{4n^2+8n+4}$$

$$= \frac{6+\frac{4}{n}}{4+\frac{8}{n}+\frac{4}{n^2}}$$

$$\therefore \quad \underset{n\to\infty}{\text{lt.}}\; n\left(\frac{u_n}{u_{n+1}} - 1\right) = \frac{6}{4} = \frac{3}{2} > 1$$

$\therefore$ Σu_n is convergent when $x^2 = 1$.

Ex. 11. *Test the series*

$$\Sigma \frac{1\cdot 2\cdot 3 \ldots n}{4\cdot 7\cdot 10 \ldots (3n+1)} x^n$$

Sol. Let the given series be denoted by Σu_n. Then,

$$u_n = \frac{1\cdot 2\cdot 3 \ldots n}{4\cdot 7\cdot 10 \ldots (3n+1)} x^n$$

$$\therefore \quad u_{n+1} = \frac{1\cdot 2\cdot 3 \ldots n(n+1)}{4\cdot 7\cdot 10 \ldots (3n+1)(3n+4)} x^{n+1}$$

$$\therefore \quad \frac{u_n}{u_{n+1}} = \frac{3n+4}{n+1}\,\frac{1}{x}$$

$$= \frac{3+\frac{4}{n}}{1+\frac{1}{n}}\,\frac{1}{x}$$

$$\therefore \quad \underset{n\to\infty}{\text{lt.}} \frac{u_n}{u_{n+1}} = \frac{3}{x}$$

Hence by ratio test Σu_n is

convergent if $\frac{3}{x} > 1$ i.e. $x < 3$

divergent if $\frac{3}{x} < 1$ i.e. $x > 3$

and the test fails if $\frac{3}{x} = 1$ i.e. $x = 3$

When $x = 3$,

we have

$$\frac{u_n}{u_{n+1}} = \frac{3n+4}{n+1} \cdot \frac{1}{3} = \frac{3n+4}{3n+3}$$

Now apply Raabe's test, we have

$$\frac{u_n}{u_{n+1}} - 1 = \frac{3n+4}{3n+3} - 1 = \frac{1}{3n+3}$$

$$\therefore \quad n\left(\frac{u_n}{u_{n+1}} - 1\right) = \frac{n}{3n+3} = \frac{1}{1+\frac{3}{n}}$$

$$\therefore \quad \underset{n\to\infty}{\text{lt.}} \left(\frac{u_n}{u_{n+1}} - 1\right) = \frac{1}{3} < 1$$

Hence Σu_n is divergent when $x = 3$.

Ex. 12. *Test the series*

$$x \log x + x^2 \log (2x) + x^3 \log (3x) + \ldots - \ldots + x^n \log (nx) + \ldots$$

Sol. Here,

$$u_n = x^n \log (nx)$$

$$u_{n+1} = x^{n+1} \log \{(n+1)x\}$$

$$\therefore \quad \frac{u_n}{u_{n+1}} = \frac{\log(nx)}{\log\{(n+1)x\}} \frac{1}{x}$$

$$= \frac{\log(nx)}{\log(nx+x)} \frac{1}{x}$$

$$= \frac{\log nx}{\log\left\{nx\left(1+\frac{1}{n}\right)\right\}} \frac{1}{x}$$

$$= \frac{\log nx}{\log\ nx + \log\left(1+\frac{1}{n}\right)} \frac{1}{x}$$

$$= \frac{\log nx}{\log\ nx + \frac{1}{n} - \frac{1}{2n^2} + \frac{1}{3n^3} - \ldots} \frac{1}{x}$$

$$= \frac{1}{1+\frac{1}{n\ \log\ nx} - \frac{1}{2n^2 \log nx} + \ldots} \frac{1}{x}$$

$$\therefore \quad \underset{n\to\infty}{\text{lt.}} \quad \frac{u_n}{u_{n+1}} = \frac{1}{x}$$

Hence by ratio test the given series is

convergent if $\frac{1}{x} > 1$ i.e. $x < 1$

divergent if $\frac{1}{x} < 1$ i.e. $x > 1$

and the test fails if $\frac{1}{x} = 1$ i.e. $x = 1$

When $x = 1$,

we have

$$\frac{u_n}{u_{n+1}} = \frac{1}{1+\frac{1}{n\ \log\ n} - \frac{1}{2n^2 \log n} + \ldots}$$

Now apply Raabe's test, we have

$$\frac{u_n}{u_{n+1}} - 1 = \frac{-\frac{1}{n\ \log\ n} + \frac{1}{2n^2 \log n} - \ldots}{1+\frac{1}{n\ \log\ n} - \frac{1}{2n^2 \log n} + \ldots}$$

$$\therefore \quad n\left(\frac{u_n}{u_{n+1}}-1\right)=\frac{n\left(-\frac{1}{n\log n}+\frac{1}{2n\log n}-\ldots\right)}{1+\frac{1}{n\log n}-\frac{1}{2n^2\log n}+\ldots}$$

$$\therefore \quad \underset{n\to\infty}{\text{lt.}}\ u\left(\frac{u_n}{u_{n+1}}-1\right)=0<1$$

Hence the given series is divergent when $x = 1$.

Ex. 13. *Test the series*

$$x^2(\log 2)^q + x^2(\log 3)^q + x^4(\log 4)^q + \ldots$$

Sol. Let us introduce the term $x(\log 1)^q$ which will not affect the series as $\log 1 = 0$, we have the given series as,

$$x(\log 1)^q + x^2(\log 2)^q + x^3(\log 3)^q + x^4(\log 4)^q + \ldots$$

Let the series now be denoted by Σu_n. Then,

$$u_n = x^n(\log n)^q$$

$$\therefore \quad u_{n+1} = x^{n+1}\{\log(n+1)\}^q$$

$$\therefore \quad \frac{u_n}{u_{n+1}} = \left[\frac{\log n}{\log(n+1)}\right]^q \frac{1}{x}$$

$$= \left[\frac{n}{\log\left\{n\left(1+\frac{1}{n}\right)\right\}}\right]^q \frac{1}{x}$$

$$= \left[\frac{\log n}{\log n+\log\left(1+\frac{1}{n}\right)}\right]^q \frac{1}{x}$$

$$= \left[\frac{\log n}{\log n+\frac{1}{n}-\frac{1}{2n^2}+\ldots}\right]^q \frac{1}{x}$$

$$= \left[\frac{1}{1+\frac{1}{n\log n}-\frac{1}{2n^2\log n}+\ldots}\right]^q \frac{1}{x}$$

$$\therefore \quad \underset{n\to\infty}{\text{lt.}} \frac{u_n}{u_{n+1}} = \frac{1}{x}$$

Hence by ratio test, the given series is

convergent if $\frac{1}{x} > 1$ i.e. $x < 1$

divergent if $\frac{1}{x} < 1$ i.e. $x > 1$

and the test fails if $\frac{1}{x} = 1$ i.e. $x = 1$

When $x = 1$, we have

$$\frac{u_n}{u_{n+1}} = \left[\frac{1}{1+\frac{1}{n\log n}-\frac{1}{2n^2\log n}+\ldots}\right]^q$$

$$= \left[1+\frac{1}{n\log n}-\frac{1}{2n^2\log n}+\ldots\right]^{-q}$$

$$= \left[1-q\left(\frac{1}{n\log n}-\frac{1}{2n^2\log n}+\ldots\right) + \frac{(-q)(-q-1)}{2!}\left(\frac{1}{n\log n}-\frac{1}{2n^2\log n}+\ldots\right)^2+\ldots\right]$$

$$= 1-\frac{q}{n\log n}+\frac{1}{2n^2}\left[\frac{q}{\log n}+\frac{q(q+1)}{(\log n)^2}\right]+\ldots$$

Now apply Raabe's test, we have

$$\frac{u_n}{u_{n+1}}-1 = -\frac{q}{n\log n}+\frac{1}{2n^2}\left[\frac{q}{\log n}+\frac{q(q+1)}{(\log n)^2}\right]+\ldots$$

$$\therefore \quad n\left(\frac{u_n}{u_{n+1}}-1\right) = -\frac{q}{\log n}+\frac{1}{2n}\left[\frac{q}{\log n}+\frac{q(q+1)}{(\log n)^2}\right]+\ldots$$

$$\therefore \quad \underset{n\to\infty}{\text{lt.}}\; n\left(\frac{u_n}{u_{n+1}}-1\right) = 0 < 1$$

Hence the given series is divergent when $x = 1$.

Ex. 14. *Test the series*

$$1+a+\frac{a(a+1)}{1\cdot 2}+\frac{a(a+1)(a+2)}{1\cdot 2\cdot 3}+....$$

Sol. Neglecting the first term, let the given series be denoted by Σu_n. Then,

$$u_n = \frac{a(a+1)(a+2)...(a+n-1)}{1\cdot 2\cdot 3...n}+....$$

$$\therefore \quad u_{n+1} = \frac{a(a+1)(a+2)...(a+n-1)(a+n)}{1\cdot 2\cdot 3...n(n+1)}+....$$

$$\therefore \quad \frac{u_n}{u_{n+1}} = \frac{n+1}{a+n} = \frac{1+\frac{1}{n}}{\frac{a}{n}+1}$$

$$\therefore \quad \underset{n\to\infty}{\text{lt.}}\; \frac{u_n}{u_{n+1}} = 1$$

Hence ratio test fails.

Let us apply now Raabe's test, we have

$$\frac{u_n}{u_{n+1}}-1 = \frac{n+1}{a+n}-1 = \frac{1-a}{a+n}$$

$$\therefore \quad n\left(\frac{u_n}{u_{n+1}}-1\right) = \frac{n(1-a)}{a+n} = \frac{1-a}{\frac{a}{n}+1}$$

$$\therefore \quad \underset{n\to\infty}{\text{lt.}}\; n\left(\frac{u_n}{u_{n+1}}-1\right) = 1-a$$

Hence by Raabe's test, Σu_n is

convergent if $1-a > 1$ i.e. $a < 0$,

divergent if $1-a < 1$ i.e. $a > 0$

and the test fails when $1-a=1$ i.e. when $a=0$

Now when $a=0$, the given series is

$$1+0+0+0+\ldots$$

$$S_n = \text{Sum of } n \text{ terms of the series}$$
$$= 1$$

$$\therefore \quad \underset{n\to\infty}{\text{lt.}} S_n = 1 \text{ which is finite.}$$

Hence the given series is convergent when $a=0$

Hence the given series is convergent if $a \le 0$ and divergent if $a > 0$.

Ex. 15. *Test for convergence the series whose n^{th} term is*

$$\frac{a^n}{a^n+x^n} \ (a \neq 0)$$

Sol. Here,

$$u_n = \frac{a^n}{a^n+x^n}$$

$$u_{n+1} = \frac{a^{n+1}}{a^{n+1}+x^{n+1}}$$

$$\frac{u_n}{u_{n+1}} = \frac{a^{n+1}+x^{n+1}}{a^n+x^n}\cdot\frac{1}{a}$$

Case I. When $x > a$, then

$$\frac{u_n}{u_{n+1}} = \frac{\left(\frac{a}{x}\right)^{n+1}+1}{\left(\frac{a}{x}\right)^{n}+1}\cdot\frac{x}{a}$$

Now as $x > a$

$$\frac{a}{x} < 1$$

$$\therefore \quad \underset{n\to\infty}{\text{lt.}} \left(\frac{a}{x}\right)^n = 0$$

$$\therefore \quad \underset{n\to\infty}{\text{lt.}} \frac{u_n}{u_{n+1}} = \frac{x}{a} > 1 \qquad (\because x > a)$$

Hence by ratio test the given series is convergent when $x > a$.

Case II. When $x < a$, then

$$\frac{u_n}{u_{n+1}} = \frac{1+\left(\frac{x}{a}\right)^{n+1}}{1+\left(\frac{x}{a}\right)^{n}}$$

Now as $x < a$

$$\frac{a}{x} < 1$$

$$\therefore \quad \underset{n\to\infty}{\text{lt.}} \left(\frac{x}{a}\right)^n = 0$$

$$\therefore \quad \underset{n\to\infty}{\text{lt.}} \frac{u_n}{u_{n+1}} = 1$$

Hence ratio test fails in this case.

But we have in this case,

$$\frac{u_n}{u_{n+1}} = \frac{1+y^{n+1}}{1+y^n}$$

where $y = \frac{x}{a} < 1$

$$\therefore \quad n\left(\frac{u_n}{u_{n+1}}-1\right) = n\left(\frac{1+y^{n+1}}{1+y^n}-1\right)$$

$$= n\left(\frac{y^{n+1}-y^n}{1+y^n}\right)^n$$

$$= \frac{n(y-1)}{1+y^{-n}}$$

$$\therefore \quad \underset{n\to\infty}{\text{lt.}} \; n\left(\frac{u_n}{u_{n+1}}-1\right) \text{ is of the form } \frac{\infty}{\infty}$$

as $y < 1$ i.e. $y^{-n} \to \infty$

as $n \to \infty$

So applying De L' Hospital's rule, we have

$$\underset{n\to\infty}{\text{lt.}}\; n\left(\frac{u_n}{u_{n+1}}-1\right)=\underset{n\to\infty}{\text{lt.}}\;\frac{y-1}{-y^{-n}\log y}$$

[Differentiating the numerator and denominator simultaneously and separately w.r.t.n]

$$=\underset{n\to\infty}{\text{lt.}}\;\frac{(1-y)y^n}{\log y}$$

$$=0 \qquad \Big|\text{as } y^n\to 0 \text{ as } n\to\infty$$

$$<1$$

Hence by Raabe's test, the given series is divergent when $x < a$.

Case III. When $x = a$, then, the series is

$$\frac{1}{2}+\frac{1}{2}+\frac{1}{2}+\ldots$$

which is clearly divergent as each term is +ve and greater than a fixed finite quantity $\frac{1}{4}$.

Hence the given series is convergent if $x > a$ and divergent if $x \le a$.

Illustrative Examples

(Based on Logarithmic Test)

Ex. 1. *Test the series*

$$x+\frac{2^2x^2}{2!}+\frac{3^3x^3}{3!}+\frac{4^4x^4}{4!}+\frac{5^5x^5}{5!}+\ldots.x>0$$

Sol. Let the given series be denoted by Σu_n. Then,

$$u_n=\frac{n^nx^n}{n!}$$

$$\therefore\quad u_{n+1}=\frac{(n+1)^{n+1}x^{n+1}}{n+1!}$$

$$\frac{u_n}{u_{n+1}}=\frac{n^n}{n!}\cdot\frac{n+1!}{(n+1)^{n+1}}\cdot\frac{1}{x}$$

$$= \frac{n^n}{(n+1)^n} \frac{1}{x}$$

$$= \frac{1}{\left(1+\frac{1}{n}\right)^n x}$$

$$\therefore \quad \underset{n \to \infty}{\text{lt.}} \frac{u_n}{u_{n+1}} = \frac{1}{ex}$$

Hence by ratio test, Σu_n is

convergent if $\frac{1}{ex} > 1$ i.e. $x < \frac{1}{e}$

divergent if $\frac{1}{ex} < 1$ i.e. $x > \frac{1}{e}$

and the test fails if $\frac{1}{ex} = 1$ i.e. $x = \frac{1}{e}$

When $\quad x = \frac{1}{e}$, then,

$$\frac{u_n}{u_{n+1}} = \frac{e}{\left(1+\frac{1}{n}\right)^n}$$

Now apply logarithmic test, we have,

$$\log \frac{u_n}{u_{n+1}} = \log \left\{ \frac{e}{\left(1+\frac{1}{n}\right)^n} \right\}$$

$$= \log e - \log \left(1+\frac{1}{n}\right)^n$$

$$= 1 - n \log \left(1+\frac{1}{n}\right)$$

$$= 1 - n \left[\frac{1}{n} - \frac{1}{2n^2} + \frac{1}{3n^3} - \ldots\right]$$

$$= 1 - \left[1 - \frac{1}{2n} + \frac{1}{3n^2} - \ldots\right]$$

$$= \frac{1}{2n} - \frac{1}{3n^2} + \ldots$$

$$n \log \frac{u_n}{u_{n+1}} = \frac{1}{2} - \frac{n}{3n} + \ldots$$

$$\therefore \quad \underset{n\to\infty}{\text{lt.}} \quad n \log \frac{u_n}{u_{n+1}} = \frac{1}{2} < 1$$

Hence Σu_n is divergent when $x = \frac{1}{e}$.

Ex. 2. *Test the series*

$$1 + \frac{x}{2} + \frac{2\,!}{3^2} x^2 + \frac{3\,!}{4^3} x^3 + \ldots$$

Sol. Leaving the first term, let the given series be denoted by Σu_n. Then,

$$u_n = \frac{n\,!\,x^n}{(n+1)^n}$$

$$u_{n+1} = \frac{n+1\,!\,x^{n+1}}{(n+2)^{n+1}}$$

$$\therefore \quad \frac{u_n}{u_{n+1}} = \frac{n\,!}{(n+1)^n} \frac{(n+2)^{n+1}}{n+1\,!} \frac{1}{x}$$

$$= \left(\frac{n+2}{n+1}\right)^{n+1} \frac{1}{x}$$

$$= \left(\frac{n+2}{n+1}\right)^{n} \left(\frac{n+2}{n+1}\right) \frac{1}{x}$$

$$= \frac{\left(1+\frac{2}{n}\right)^n}{\left(1+\frac{1}{n}\right)^n} \frac{1+\frac{2}{n}}{1\,\frac{1}{n}} \frac{1}{x}$$

$$\therefore \quad \underset{n\to\infty}{\text{lt.}} \quad \frac{u_n}{u_{n+1}} = \frac{e^2}{e} \frac{1}{x} = \frac{e}{x}$$

Hence by ratio test, Σu_n is

convergent if $\frac{e}{x} > 1$ i.e. $x < e$

is divergent if $\frac{e}{x} < 1$ i.e. $x > e$

and the test fails if $\frac{e}{x} = 1$ i.e. $x = e$

When $\quad x = e$

$$\frac{u_n}{u_{n+1}} = \frac{\left(1+\frac{2}{n}\right)^{n+1}}{\left(1+\frac{1}{n}\right)^{n+1}} \frac{1}{e}$$

Now apply logarithmic test, we have

$$\log \frac{u_n}{u_{n+1}} = (n+1)\left[\log\left(1+\frac{2}{n}\right) - \log\left(1+\frac{1}{n}\right)\right] - \log e$$

$$= (n+1)\left[\left\{\frac{2}{n} - \frac{1}{2}\left(\frac{2}{n}\right)^2 + \frac{1}{3}\left(\frac{2}{n}\right)^3 - \ldots\right\} - \left\{\frac{1}{n} - \frac{1}{2}\left(\frac{1}{n}\right)^2 + \frac{1}{3}\left(\frac{1}{n}\right)^3 - \ldots\right\}\right] - 1$$

$$= (n+1)\left[\left\{\frac{2}{n} - \frac{2}{n^2} + \frac{8}{3n^3} - \ldots\right\} - \left\{\frac{1}{n} - \frac{1}{2n^2} + \frac{1}{3n^3} - \ldots\right\}\right] - 1$$

$$= (n+1)\left[\frac{1}{n} - \frac{3}{2n^2} + \frac{7}{3n^3} - \ldots\right] - 1$$

$$= \left[1 + \frac{1}{n}\left(1 - \frac{3}{2}\right) + \frac{1}{n^2}\left(\frac{7}{3} - \frac{3}{2}\right) + \ldots\right] - 1$$

$$= -\frac{1}{2n} + \frac{5}{6n^2} + \ldots$$

$$\therefore \quad n \log \frac{u_n}{u_{n+1}} = -\frac{1}{2} + \frac{5}{6n} + \ldots.$$

$$\therefore \quad \underset{n\to\infty}{\text{lt.}}\ n \log \frac{u_n}{u_{n+1}} = -\frac{1}{2} < 1$$

Hence Σu_n is divergent, when $x = e$

Ex. 3. *Test the series*

$$1+\frac{2x}{2!}+\frac{3^2x^2}{3!}+\frac{4^3x^3}{4!}+\ldots,\ x>0$$

Sol. Let the given series be denoted by Σu_n. Then,

$$u_n=\frac{n^{n-1}x^{n-1}}{n!}$$

$$\therefore \qquad u_{n+1}=\frac{(n+1)^n x^n}{n+1!}$$

$$\therefore \qquad \frac{u_n}{u_{n+1}}=\frac{n^{n-1}}{(n+1)^n}\frac{n+1!}{n!}\frac{1}{x}$$

$$=\left(\frac{n}{n+1}\right)^{n-1}\frac{1}{x}$$

$$=\left(\frac{n}{n+1}\right)^{n}\left(\frac{n}{n+1}\right)^{-1}\frac{1}{x}$$

$$=\frac{1}{\left(1+\frac{1}{n}\right)^n}\left(\frac{n+1}{n}\right)\frac{1}{x}$$

$$=\frac{1+\frac{1}{n}}{\left(1+\frac{1}{n}\right)^n}\frac{1}{x}$$

$$\therefore \qquad \underset{n\to\infty}{\text{lt.}}\ \frac{u_n}{u_{n+1}}=\frac{1}{ex}$$

Hence by ratio test, Σu_n is

convergent if $\frac{1}{ex}>1$ i.e. $x<\frac{1}{e}$

divergent if $\frac{1}{ex}<1$ i.e. $x>\frac{1}{e}$

and the test fails if $\frac{1}{ex}=1$ i.e. $x=\frac{1}{e}$

When $\qquad x = \dfrac{1}{e}$, then

$$\frac{u_n}{u_{n+1}} = \frac{e}{\left(1+\dfrac{1}{n}\right)^{n-1}}$$

$$\log \frac{u_n}{u_{n+1}} = \log e - (n-1)\log\left(1+\frac{1}{n}\right)$$

$$= 1 - (n-1)\left[\frac{1}{n} - \frac{1}{2n^2} + \frac{1}{3n^3} - \ldots\right]$$

$$= 1 - \left[1 + \frac{1}{n}\left(-\frac{1}{2} - 1\right) + \frac{1}{n^2}\left(\frac{1}{3} + \frac{1}{2}\right) + \ldots.\right]$$

$$= \frac{3}{2n} - \frac{5}{6n^2} + \ldots.$$

$$\therefore \quad n \log \frac{u_n}{u_{n+1}} = \frac{3}{2} - \frac{5}{6n} + \ldots.$$

$$\therefore \quad \lim_{n\to\infty} n \log \frac{u_n}{u_{n+1}} = \frac{3}{2} > 1$$

Hence Σu_n is convergent when $x = \dfrac{1}{e}$.

Ex. 4. *Test the series*

$$\frac{a+x}{1!} + \frac{(a+2x)^2}{2!} + \frac{(a+3x)^3}{3!} + \ldots.$$

Sol. Let the given series be denoted by Σu_n.

Then,

$$u_n = \frac{(a+nx)^n}{n!}$$

Compare the series Σu_n with the auxiliary series

$$\Sigma v_n = \Sigma \frac{n^n x^n}{n!}$$

$$\therefore \quad \frac{u_n}{v_n} = \frac{(a+nx)^n}{n\,!} \; \frac{n\,!}{n^n x^n}$$

$$= \left(\frac{a+nx}{nx} \right)^n$$

$$= \left(1 + \frac{a}{nx} \right)^n$$

$$= \left[\left(1 + \frac{a}{nx} \right)^{nx/a} \right]^{a/x}$$

$$\therefore \quad \underset{n\to\infty}{\text{lt.}} \quad = \underset{n\to\infty}{\text{lt.}} \left[\left(1 + \frac{a}{nx} \right)^{nx/a} \right]^{a/x} = e^{a/x}$$

which is a fixed finite non-zero quantity as a and x are finite.

Hence comparison test can be applied. Thus Σu_n and Σv_n are either both convergent or both divergent i.e. the condition under which Σv_n is convergent or divergent will hold good for the convergence or divergence of Σu_n.

But we have already discussed the convergence or divergence of Σv_n in Ex. 1.

Hence by Ex. 1, Σu_n will be

convergent if $x < \dfrac{1}{e}$

and divergent if $x \geq \dfrac{1}{e}$.

Ex. 5. *Test the series*

$$\Sigma \left(\frac{nx}{n+1} \right)^n$$

Sol. Let the given series be denoted by Σu_n. Then,

$$u_n = \frac{n^n x^n}{(n+1)^n}$$

$$\therefore \quad u_{n+1} = \frac{(n+1)^{n+1} x^{n+1}}{(n+2)^{n+1}}$$

$$\therefore \quad \frac{u_n}{u_{n+1}} = \frac{n^n}{(n+1)^n} \cdot \frac{(n+2)^{n+1}}{(n+1)^{n+1}} \cdot \frac{1}{x}$$

$$= \frac{1}{\left(1+\frac{1}{n}\right)^n} \left(1+\frac{1}{n+1}\right)^{n+1} \frac{1}{x}$$

$$\therefore \quad \underset{n\to\infty}{\text{lt.}} \frac{u_n}{u_{n+1}} = \frac{1}{x}$$

as $\underset{n\to\infty}{\text{lt.}} \left(1+\frac{1}{n}\right)^n = e$

and $\underset{n\to\infty}{\text{lt.}} \left(1+\frac{1}{n+1}\right)^{n+1} = e$

Hence by ratio test Σu_n is

convergent if $\frac{1}{x} > 1$ i.e. $x < 1$

divergent if $\frac{1}{x} < 1$ i.e. $x > 1$

and the test fails if $\frac{1}{x} = 1$ i.e. $x = 1$

When $x = 1$,

we have,

$$\frac{u_n}{u_{n+1}} = \frac{1}{\left(1+\frac{1}{n}\right)^n} \left(1+\frac{1}{n+1}\right)^{n+1}$$

Now apply Logarithmic test, we have

$$n \log \frac{u_n}{u_{n+1}} = n \left[(n+1) \log \left(1+\frac{1}{n+1}\right) - n \log \left(1+\frac{1}{n}\right)\right]$$

$$n\left[(n+1)\left\{\frac{1}{n+1} - \frac{1}{2(n+1)^2} + \frac{1}{3(n+1)^3} - \dots\right\} - n\left\{\frac{1}{n} - \frac{1}{2n^2} + \frac{1}{3n^3} - \dots\right\}\right]$$

$$\therefore \quad \underset{n\to\infty}{\text{lt.}}\ n\log\frac{u_n}{u_{n+1}} = \underset{n\to\infty}{\text{lt.}}\ n\left[\frac{1}{2n}-\frac{1}{2(n+1)}+\frac{1}{3(n+1)^2}-\frac{1}{3n^2}+\ldots\right]$$

$$= 0 < 1.$$

Hence Σu_n is divergent when $x = 1$.

Illustrative Examples

(Based on Cauchy's Condensation Test)

Ex. 1. *Test the series*

$$\frac{1}{(\log 2)^p}+\frac{1}{(\log 3)^p}+\frac{1}{(\log 4)^p}+\ldots+\frac{1}{(\log n)^p}+\ldots$$

Sol. Applying Cauchy's condensation test, let

$$f(n) = \frac{1}{(\log n)^p}$$

Clearly $f(n) \to 0$ as $n \to \infty$.

Now,

$$m^n f(m^n) = m^n \frac{1}{(\log m^n)^p} \qquad \Big|\text{ where } m > 1$$

$$= m^n \frac{1}{(n \log m)^p}$$

$$= m^n \frac{1}{(n^p \log m)^p}$$

$$= \frac{1}{(\log m)^p}\cdot\frac{m^n}{n^p}$$

Again, let $\Sigma v_n = \Sigma \dfrac{1}{(\log m)^p}\cdot\dfrac{m^n}{n^p}$

then $$v_n = \frac{1}{(\log m)^p}\cdot\frac{m^n}{n^p}$$

$$v_{n+1} = \frac{1}{(\log m)^p}\cdot\frac{m^{n+1}}{(n+1)^p}$$

$$\therefore \quad \frac{v_n}{v_{n+1}} = \frac{m^n}{m^{n+1}}\frac{(n+1)^p}{n^p}$$

$$= \left(1+\frac{1}{n}\right)^p \cdot \frac{1}{m}$$

$$\therefore \quad \underset{n\to\infty}{\text{lt.}} \frac{v_n}{v_{n+1}} = \frac{1}{m} < 1 \text{ as } m > 1$$

Hence by ratio test, Σv_n is divergent.

i.e. $\Sigma \dfrac{1}{(\log m)^p} \dfrac{m^n}{n^p}$ is divergent

Hence by Cauchy's condensation test,

$$f(n) = \Sigma \frac{1}{(\log n)^p} \text{ is also divergent.}$$

Ex. 2. *Test the series*

$$\Sigma\left(\frac{\log n}{n}\right)$$

Sol. The first term of the series can be taken as

$$\frac{\log 1}{1} \text{ as } \log 1 = 0$$

Applying Cauchy's condensation test,

let $\quad f(n) = \dfrac{\log n}{n}$

Clearly $f(n) \to 0 \quad$ as $n \to \infty$

$$\therefore \quad \underset{n\to\infty}{\text{lt.}} \frac{\log n}{n} = 0$$

Now, $m^n f(m^n) = m^n \dfrac{\log m^n}{m^n}$

$= \log m^n = n \log m$ where $m > 1$

Let $\quad \Sigma v_n = \Sigma n \log m$

$= \log m \, \Sigma n = \log m \, (1 + 2 + 3 + ...)$

Clearly Σv_n is divergent.

i.e. $\Sigma n \log m$ is divergent.

Hence by Cauchy's condensation test

$$\Sigma f(n) = \Sigma \frac{\log n}{n} \text{ is also divergent.}$$

Ex. 3. *Test the series*

$$\frac{(\log 2)^2}{2^2}+\frac{(\log 3)^2}{3^2}+\frac{(\log 4)^2}{4^2}+\ldots+\frac{(\log n)^n}{n^2}+\ldots$$

Sol. The first term of the series can be taken as

$$\frac{(\log 1)^2}{1^2} \text{ as } \log 1 = 0$$

Applying Cauchy's condensation test,

$$\text{let} \quad f(n)=\frac{(\log n)^2}{n^2}$$

Clearly $f(n)\to 0$ as $n\to\infty$

$$\text{Now, } m^n f(m^n) = m^n \frac{(\log m^n)^2}{(m^n)^2} \text{ where } m > 1$$

$$=\frac{(n\log m)^2}{m^n}$$

$$=\frac{n^2(\log m)^2}{m^n}$$

$$\text{Let} \qquad \Sigma v_n = \Sigma\frac{n^2(\log m)^2}{m^n}$$

$$\text{then,} \qquad v_n=\frac{n^2(\log m)^2}{m^n}$$

$$\therefore \qquad v_{n+1}=\frac{(n+1)^2(\log m)^2}{m^{n+1}}$$

$$\therefore \qquad \frac{v_n}{v_{n+1}}=\frac{n^2}{(n+1)^2}\cdot m$$

$$=\frac{1}{\left(1+\frac{1}{n}\right)^2}\cdot m$$

$$\lim_{n\to\infty}\frac{v_n}{v_{n+1}}=m>1$$

e by ratio test, Σv_n is convergent i.e.

$$\Sigma \frac{n^2 (\log m)^2}{m^n}$$

or $\Sigma m^n f(m^n)$ is convergent.

Hence by Cauchy's condensation test $\Sigma f(n)$ i.e. the given series is also convergent.

Ex. 4. *Test the series*

$$\Sigma \frac{n^{-1}}{\sqrt{\log n}}$$

Sol. Let the given series be Σu_n then

$$\Sigma u_n = \Sigma \frac{n^{-1}}{\sqrt{\log n}}$$

$$= \Sigma \frac{1}{n(\log n)^{1/2}}$$

which is of the form $\Sigma \dfrac{1}{n(\log n)^p}$ where $p = \dfrac{1}{2}$ i.e. < 1.

Hence Σu_n is divergent by Art.1.24.

Ex. 5. *Test the series*

$$\Sigma \frac{1}{n \log n}$$

Sol. Proceed as in Ex. 4, as $p = 1$.

By Art. 1.24, the given series is divergent.

Illustrative Examples

(Based on De Morgan and Bertrand's Test)

Ex. 1. *Test the series*

$$1 + \frac{2^2}{3^2} + \frac{2^2 \cdot 4^2}{3^2 \cdot 5^2} + \frac{2^2 \cdot 4^2 \cdot 6^2}{3^2 \cdot 5^2 \cdot 7^2} + \ldots$$

Sol. Leaving the first term, let the given series be denoted by Σu_n. Then,

$$u_n = \frac{2^2 \cdot 4^2 \cdot 6^2 \ldots (2n)^2}{3^2 \cdot 5^2 \cdot 7^2 \ldots (2n+1)^2}$$

$$\therefore \quad u_{n+1} = \frac{2^2 \cdot 4^2 \cdot 6^2 \ldots (2n)^2 (2n+2)^2}{3^2 \cdot 5^2 \cdot 7^2 \ldots (2n+1)^2 (2n+3)^2}$$

$$\therefore \quad \frac{u_n}{u_{n+1}} = \left(\frac{2n+3}{2n+2}\right)^2$$

$$= \left[\frac{2+\frac{3}{n}}{2+\frac{2}{n}}\right]^2$$

$$\therefore \quad \lim_{n\to\infty} \frac{u_n}{u_{n+1}} = 1$$

Hence ratio test fails.

Now applying Raabe's test, we have

$$\frac{u_n}{u_{n+1}} = \frac{4n^2+12n+9}{4n^2+8n+4}$$

$$\frac{u_n}{u_{n+1}} - 1 = \frac{4n+5}{4n^2+8n+4}$$

$$n\left(\frac{u_n}{u_{n+1}} - 1\right) = \frac{4n^2+5n}{4n^2+8n+4}$$

$$= \frac{4+\frac{5}{n}}{4+\frac{8}{n}+\frac{4}{n^2}}$$

$$\therefore \quad \lim_{n\to\infty} n\left(\frac{u_n}{u_{n+1}} - 1\right) = 1$$

Hence Raabe's test fails.

Now applying De Morgan and Bertrand's Test, we have

$$n\left(\frac{u_n}{u_{n+1}} - 1\right) = \frac{4n^2+5n}{4n^2+8n+4}$$

$$\therefore \quad n\left(\frac{u_n}{u_{n+1}} - 1\right) - 1 = \frac{-3n-4}{4n^2+8n+4}$$

$$\therefore \quad \log n\left[n\left(\frac{u_n}{u_{n+1}} - 1\right) - 1\right] = \frac{(-3n-4)\log n}{4n^2+8n+4}$$

$$= \frac{-3n^2 - 4n}{4n^2 + 8n + 4} \cdot \frac{\log n}{n}$$

$$= \frac{-3 - \frac{4}{n}}{4 + \frac{8}{n} + \frac{4}{n^2}} \cdot \frac{\log n}{n}$$

$$\therefore \quad \underset{n \to \infty}{\text{lt.}} \log n \left[n \left(\frac{u_n}{u_{n+1}} - 1 \right) - 1 \right]$$

$$= \underset{n \to \infty}{\text{lt.}} \frac{-3 - \frac{4}{n}}{4 + \frac{8}{n} + \frac{4}{n^2}} \cdot \frac{\log n}{n}$$

$$= -\frac{3}{4} . 0$$

$$\left(\because \underset{n \to \infty}{\text{lt.}} \frac{\log n}{n} = 0 \right)$$

$$= 0 < 1$$

Hence the given series is divergent.

Ex. 2. *Test the series*

$$\Sigma \frac{2^2 \cdot 4^2 ... (2n-2)^2}{3^2 \cdot 5^2 ... (2n-1)^2} + ...$$

Sol. Let the given series be denoted by Σu_n. Then, we have

$$u_n = \frac{2^2 \cdot 4^2 ... (2n-2)^2}{3^2 \cdot 5^2 ... (2n-1)^2}$$

$$u_{n+1} = \frac{2^2 \cdot 4^2 ... (2n-2)^2 \, (2n)^2}{3^2 \cdot 3^2 ... (2n-1)^2 \, (2n+1)^2}$$

$$\frac{u_n}{u_{n+1}} = \left(\frac{2n+1}{2n} \right)^2$$

$$= \left(1 + \frac{1}{2n} \right)^2$$

$$\therefore \quad \underset{n \to \infty}{\text{lt.}} \frac{u_n}{u_{n+1}} = 1$$

Hence ratio test fails.

Now apply Raabe's test, we have

$$\frac{u_n}{u_{n+1}} = \frac{4n^2+4n+1}{4n^2}$$

$$\therefore \quad \frac{u_n}{u_{n+1}} - 1 = \frac{4n+1}{4n^2}$$

$$\therefore \quad n\left(\frac{u_n}{u_{n+1}} - 1\right) = \frac{4n^2+n}{4n^2} = 1 + \frac{1}{4n}$$

$$\therefore \quad \underset{n\to\infty}{\text{lt.}}\ n\left(\frac{u_n}{u_{n+1}} - 1\right) = 1$$

Hence Raabe's test fails.

Now apply De Morgan and Bertrand's test, we have

$$n\left(\frac{u_n}{u_{n+1}} - 1\right) = \frac{4n^2+n}{4n^2}$$

$$\therefore \quad n\left(\frac{u_n}{u_{n+1}} - 1\right) - 1 = \frac{n}{4n^2} = \frac{1}{4n}$$

$$\therefore \quad \log n\left[n\left(\frac{u_n}{u_{n+1}} - 1\right) - 1\right] = \frac{1}{4}.\frac{\log n}{n}$$

$$\therefore \quad \underset{n\to\infty}{\text{lt.}}\ \log n\left[n\left(\frac{u_n}{u_{n+1}} - 1\right) - 1\right]$$

$$= \underset{n\to\infty}{\text{lt.}}\ \frac{1}{4}\frac{\log n}{n}$$

$$= \frac{1}{4}\ .\ 0$$

$$= 0 < 1.$$

Hence the given series is divergent.

Ex. 3. *Test the series*

$$\frac{1^2}{2^2}+\frac{1^2\cdot 3^2}{2^2\cdot 4^2}\cdot x+\frac{1^2\cdot 3^2\cdot 5^2}{2^2\cdot 4^2\cdot 6^2}x^2+\ldots(x>0)$$

Sol. Leaving the first term, let the given series be denoted by Σu_n. Then,

$$u_n=\frac{1^2\cdot 3^2\cdot 5^2\ldots(2n+1)^2}{2^2\cdot 4^2\cdot 6^2\ldots(2n+2)^2}x^n$$

$$\therefore \quad u_{n+1}=\frac{1^2\cdot 3^2\cdot 5^2\ldots(2n+1)^2\,(2n+3)^2}{2^2\cdot 4^2\cdot 6^2\ldots(2n+2)^2\,(2n+4)^2}x^{n+1}$$

$$\therefore \quad \frac{u_n}{u_{n+1}}=\frac{(2n+4)^2}{(2n+3)^2}\frac{1}{x}$$

$$=\left[\frac{2+\frac{4}{n}}{2+\frac{3}{n}}\right]^2\frac{1}{x}$$

$$\therefore \quad \lim_{n\to\infty}\frac{u_n}{u_{n+1}}=\frac{1}{x}$$

Hence by ratio test, the given series is

convergent when $\frac{1}{x}>1$ i.e. $x<1$

divergent when $\frac{1}{x}<1$ i.e. $x>1$

and the test fails when $\frac{1}{x}=1$ i.e. $x=1$

When $x=1$, we have

$$\therefore \quad \frac{u_n}{u_{n+1}}=\left(\frac{2n+4}{2n+3}\right)^2$$

$$=\frac{4n^2+16n+16}{4n^2+12n+9}$$

Now apply Raabe's test, we have

$$\frac{u_n}{u_{n+1}}-1=\frac{4n+7}{4n^2+12n+9}$$

$$\therefore \quad n\left(\frac{u_n}{u_{n+1}}-1\right)=\frac{4n^2+7}{4n^3+12n+9}$$

$$=\frac{4+\frac{7}{n}}{4+\frac{12}{n}+\frac{9}{n^2}}$$

$$\therefore \quad \underset{n\to\infty}{\text{lt.}}\ n\left(\frac{u_n}{u_{n+1}}-1\right)=1$$

Hence Raabe's test fails.

Now applying De Morgan and Bertrand's test, we have

$$n\left(\frac{u_n}{u_{n+1}}-1\right)-1=\frac{4n^2+7}{4n^2+12n+9}-1$$

$$=\frac{-5n-9}{4n^2+12n+9}$$

$$\therefore \quad \log n\left[n\left(\frac{u_n}{u_{n+1}}-1\right)-1\right]=\frac{(-5n-9)\log n}{4n^2+12n+9}$$

$$=\frac{(5n^2+9n)}{4n^2+12n+9}\frac{\log n}{n}$$

$$=-\frac{5+\frac{9}{n}}{4+\frac{12}{n}+\frac{9}{n^2}}\frac{\log n}{n}$$

$$\therefore \quad \underset{n\to\infty}{\text{lt.}}\ \log n\left[n\left(\frac{u_n}{u_{n+1}}-1\right)-1\right]$$

$$=\underset{n\to\infty}{\text{lt.}}\ -\frac{5+\frac{9}{n}}{4+\frac{12}{n}+\frac{9}{n^2}}\left(\frac{\log n}{n}\right)$$

$$= -\frac{5}{4} \times 0$$
$$= 0 < 1$$

Hence the given series is divergent when $x = 1$.

Ex. 4. *Test the series*

$$\frac{a}{b} + \frac{a(a+d)}{b(b+d)} + \frac{a(a+d)(a+2d)}{b(b+d)(b+2d)} + ...$$

Sol. Let the given series be denoted by Σu_n. Then,

$$u_n = \frac{a(a+d)(a+2d)...(a+\overline{n-1}d)}{b(b+d)(b+2d)...(b+\overline{n-1}d)}$$

$$u_{n+1} = \frac{a(a+d)(a+2d)...(a+\overline{n-1}d)(a+nd)}{b(b+d)(b+2d)...(b+\overline{n-1}d)(b+nd)}$$

$$\frac{u_n}{u_{n+1}} = \frac{b+nd}{a+nd}$$

$$= \frac{\frac{b}{n}+d}{\frac{a}{n}+d}$$

$$\therefore \quad \underset{n\to\infty}{\text{lt.}} \frac{u_n}{u_{n+1}} = 1$$

Hence Ratio test fails.

Now apply Raabe's test, we have

$$\frac{u_n}{u_{n+1}} - 1 = \frac{b+nd}{a+nd} - 1$$

$$= \frac{b-a}{a+nd}$$

$$\therefore \quad n\left(\frac{u_n}{u_{n+1}} - 1\right) = \frac{(b-a)n}{a+nd}$$

$$= \frac{b-a}{\frac{a}{n}+d}$$

$$\therefore \quad \underset{n\to\infty}{\text{lt.}} \; n\left(\frac{u_n}{u_{n+1}}-1\right)=\frac{b-a}{d}$$

Hence Σu_n is

convergent if $\dfrac{b-a}{d} > 1$ i.e. $b-a > d$,

divergent if $\dfrac{b-a}{d} < 1$ i.e. $b-a < d$

and the test fails if $\dfrac{b-a}{d} = 1$ i.e. $b-a = d$

When $\quad b-a=d,$

Apply De Morgan and Bertrand's test,

we have

$$n\left(\frac{u_n}{u_{n+1}}-1\right)-1=\frac{(b-a)n}{a+nd}-1$$

$$=\frac{nd}{a+nd}-1$$

$$=-\frac{a}{a+nd}$$

$$\therefore \quad \log u\left[n\left(\frac{u_n}{u_{n+1}}-1\right)-1\right]=-\frac{a\log n}{a+nd}$$

$$=-\frac{a}{\frac{a}{n}+d}\,\frac{\log n}{n}$$

$$\therefore \quad \underset{n\to\infty}{\text{lt.}} \; \log n\left[u\left(\frac{u_n}{u_{n+1}}-1\right)-1\right]$$

$$= \underset{n\to\infty}{\text{lt.}} -\frac{a}{\frac{a}{n}+d}\,\frac{\log n}{n}$$

$$=-\frac{a}{d}.0$$

$$=0<1$$

Hence Σu_n is divergent when $b - a = d$

Ex. 5. *Test the series*

$$\frac{a}{b} + \frac{a(a+1)}{b(b+1)} + \frac{a(a+1)(a+2)}{b(b+1)(b+2)}$$

Sol. Proceed exactly as in Ex. 4, $(d = 1)$

The given series is convergent when $b - a > 1$ and divergent when $b - a < 1$.

Ex. 6. *Test the series*

$$1 + \frac{a(1-a)}{1^2} + \frac{(1+a)a(1-a)(2-a)}{1^2 \cdot 2^2} + \frac{(2+a)(1+a)a(1-a)(2-a)(3-a)}{1^2 \cdot 2^2 \cdot 3^2} + \ldots$$

where $0 < a < 1$

Sol. Leaving the first term, let the given series be denoted by Σu_n.
Then,

$$u_n = \frac{(n-1+a)\ldots(2+a)(1+a)a(1-a)(2-a)\ldots(n-a)}{1^2 \cdot 2^2 \cdot 3^2\, n^2}$$

$$u_{n+1} = \frac{(n+a)(n-1+a)\ldots(2+a)(1+a)a(1-a)(2-a)\ldots(n-a)(n+1-a)}{1^2 \cdot 2^2 \cdot 3^2 \ldots n^2 (n+1)^2}$$

$$\therefore \quad \frac{u_n}{u_{n+1}} = \frac{(n+1)^2}{(n+a)(n+1-a)}$$

$$= \frac{\left(1+\frac{1}{n}\right)^2}{\left(1+\frac{a}{n}\right)\left(1+\frac{1-a}{n}\right)}$$

$$\therefore \quad \underset{n\to\infty}{\text{lt.}} \frac{u_n}{u_{n+1}} = 1$$

Hence ratio test fails.

Now applying Raabe's test, we have

$$\frac{u_n}{u_{n+1}}-1=\frac{(n+1)^2}{(n+a)(n+1-a)}-1$$

$$=\frac{n^2+2n+1}{n^2-a^2+n+a}$$

$$=\frac{n+1-a+a^2}{n^2+n+a-a^2}$$

$$n\left(\frac{u_n}{u_{n+1}}-1\right)=\frac{n^2+(1-a+a^2)n}{n^2+n+a-a^2}$$

$$=\frac{1+\dfrac{1-a+a^2}{n}}{1+\dfrac{1}{n}+\dfrac{a-a^2}{n^2}}$$

$$\therefore \quad \underset{n\to\infty}{\text{lt.}} \; n\left(\frac{u_n}{u_{n+1}}-1\right)=1$$

Hence Raabe's test fails.

Now apply De Morgan and Bertrand's test, we have

$$n\left(\frac{u_n}{u_{n+1}}-1\right)-1=\frac{n^2+(1-a+a^2)n}{n^2+n+a-a^2}-1$$

$$=\frac{(a^2-a)n+(a^2-a)}{n^2+n+(a-a^2)}$$

$$=\frac{(a^2-a)(n+1)}{n^2+n+a-a^2}$$

$$=\frac{(a^2-a)\left(1+\dfrac{1}{n}\right)}{1+\dfrac{1}{n}+\dfrac{a-a^2}{n}}\frac{1}{n}$$

$$\therefore \quad \log n\left[n\left(\frac{u_n}{u_{n+1}}-1\right)-1\right]$$

$$=\frac{(a^2-a)\left(1+\frac{1}{n}\right)}{1+\frac{1}{n}+\frac{a-a^2}{n}}\frac{\log n}{n}$$

$$\therefore \quad \underset{n\to\infty}{\text{lt.}} \log n\left[n\left(\frac{u_n}{u_{n+1}}-1\right)-1\right]$$

$$\underset{n\to\infty}{\text{lt.}} \frac{(a^2-a)\left(1+\frac{1}{n}\right)}{1+\frac{1}{n}+\frac{a-a^2}{n}}\frac{\log n}{n}$$

$$=(a^2-a).0$$

$$=0<1.$$

Hence the given series is divergent.

Ex. 7. *Test the series*

$$\frac{(1+a)(1+b)}{1\cdot2\cdot3}+\frac{(2+a)(2+b)}{2\cdot3\cdot4}+\frac{(3+a)(3+b)}{3\cdot4\cdot5}+\ldots.$$

Sol. Let the given series be denoted by Σu_n. Then,

$$u_n=\frac{(n+a)(n+b)}{n(n+1)(n+2)}$$

$$u_{n+1}=\frac{(n+1+a)(n+1+b)}{(n+1)(n+2)(n+3)}$$

$$\therefore \quad \frac{u_n}{u_{n+1}}=\frac{(n+a)(n+b)(n+3)}{n(n+1+a)(n+1+b)}$$

$$=\frac{\left(1+\frac{a}{n}\right)\left(1+\frac{b}{n}\right)\left(1+\frac{3}{n}\right)}{\left(1+\frac{1+a}{n}\right)\left(1+\frac{1+b}{n}\right)}$$

$$\therefore \quad \underset{n\to\infty}{\text{lt.}} \frac{u_n}{u_{n+1}}=1$$

Hence ratio test fails.

Now apply Raabe's test, we have

$$\frac{u_n}{u_{n+1}}-1=\frac{(n+a)(n+b)(n+3)}{n(n+1+a)(n+1+b)}-1$$

$$=\frac{n^3+n^2(a+b+3)+n(ab+3a+3b)+3ab}{n^3+n^2(a+b+2)+n(ab+a+b+1)}-1$$

$$=\frac{n^2+n(2a+2b-1)+3ab}{n^3+n^2(a+b+2)+n(ab+a+b+1)}$$

$$\therefore \quad n\left(\frac{u_n}{u_{n+1}}-1\right)$$

$$=\frac{n^3+n^2(2a+2b-1)+3ab}{n^3+n^2(a+b+2)+n(ab+a+b+1)}$$

$$=\frac{1+\dfrac{2a+2b-1}{n}+\dfrac{3ab}{n^2}}{1+\dfrac{a+b+2}{n}+\dfrac{ab+a+b+1}{n^2}}$$

$$\therefore \quad \underset{n\to\infty}{\text{lt.}}\; n\left(\frac{u_n}{u_{n+1}}-1\right)=1$$

Hence Raabe's test fails.

Now apply De Morgan and Bertrand's test, we have

$$n\left(\frac{u_n}{u_{n+1}}-1\right)-1$$

$$=\frac{n^2(a+b-3)+n(2ab-a-b-1)}{n^3+n^2\cdot(a+b+2)+n(ab+a+b+1)}$$

$$=\frac{1}{n}\cdot\frac{a+b-3+\dfrac{2ab-a-b-1}{n}}{1+\dfrac{a+b+2}{n}+\dfrac{ab+a+b+1}{n^2}}$$

$$\therefore \quad \log n\left[n\left(\frac{u_n}{u_{n+1}}-1\right)-1\right]$$

$$=\frac{\log n}{n}\cdot\frac{a+b-3+\dfrac{2ab-a-b-1}{n}}{1+\dfrac{a+b+2}{n}+\dfrac{ab+a+b+1}{n^2}}$$

$$\therefore \quad \underset{n\to\infty}{\text{lt.}}\; n\log\left[n\left(\frac{u_n}{u_{n+1}}-1\right)-1\right]$$

$$=0.(a+b-3)$$

$$=0<1.$$

Hence the given series is divergent.

IInd Method. We have

$$u_n=\frac{(n+a)(n+b)}{n(n+1)(n+2)}$$

$$v_n=\frac{n^2}{n^3}=\frac{1}{n}$$

where v_n is n^{th} term of the auxiliary series Σu_n.

$$\frac{u_n}{v_n}=\frac{(n+a)(n+a)}{(n+1)(n+2)}$$

$$=\frac{\left(1+\dfrac{a}{n}\right)\left(1+\dfrac{b}{n}\right)}{\left(1+\dfrac{1}{n}\right)\left(1+\dfrac{2}{n}\right)}$$

$$\therefore \quad \underset{n\to\infty}{\text{lt.}}\;\frac{u_n}{v_n}=1$$

which is a fixed finite non-zero quantity.

Hence comparison test can be applied.

But $\Sigma\frac{1}{n}$ is divergent as $p=1$ by p-test.

Hence Σu_n is also divergent by comparison test.

Illustrative Examples

(Based on Special Logarithmic Test)

Ex. 1. *Test the series*

$$x + x^{1+\frac{1}{2}} + x^{1+\frac{1}{2}+\frac{1}{3}} + x^{1+\frac{1}{2}+\frac{1}{3}+\frac{1}{4}} + \ldots$$

Sol. Here,

$$u_n = x^{1+\frac{1}{2}+\frac{1}{3}+\ldots+\frac{1}{n}}$$

$$\therefore \quad u_{n+1} = x^{1+\frac{1}{2}+\frac{1}{3}+\ldots+\frac{1}{n}+\frac{1}{n+1}}$$

$$\therefore \quad \frac{u_n}{u_{n+1}} = \frac{1}{x^{\frac{1}{n+1}}} = \left(\frac{1}{x}\right)^{\frac{1}{n+1}}$$

$$\therefore \quad n \log \frac{u_n}{u_{n+1}} = n \log \left(\frac{1}{x}\right)^{\frac{1}{n+1}}$$

$$= \frac{n}{n+1} \log \left(\frac{1}{x}\right)$$

$$= \frac{1}{1+\frac{1}{n}} \log \left(\frac{1}{x}\right)$$

$$\therefore \quad \underset{n\to\infty}{\text{lt.}}\ n \log \frac{u_n}{u_{n+1}} = \log \left(\frac{1}{x}\right)$$

Hence by Logarithmic test, the given series is

convergent if $\log \left[\frac{1}{x}\right] > 1$

i.e. $\log \left[\frac{1}{x}\right] > \log e$ i.e. $\frac{1}{x} > e$ i.e. $x < \frac{1}{e}$

and divergent if $\log\left[\frac{1}{x}\right] < 1$ i.e. $x > \frac{1}{e}$

and the test fails if $\log\left[\frac{1}{x}\right] = 1$ i.e. $x = \frac{1}{e}$

Now when $x = \frac{1}{e}$, we have

$$n \log \frac{u_n}{u_{n+1}} = \frac{n}{n+1} \log (e) = \frac{n}{n+1}$$

$$\therefore \quad n \log \frac{u_n}{u_{n+1}} - 1 = \frac{n}{n+1} - 1 = \frac{-1}{n+1}$$

$$\therefore \quad \log n\left[n \log \frac{u_n}{u_{n+1}} - 1\right] = -\frac{\log n}{n+1}$$

$$\underset{n\to\infty}{\text{lt.}} \log n\left[n \log \frac{u_n}{u_{n+1}} - 1\right] = \underset{n\to\infty}{\text{lt.}} -\frac{\log n}{n} \cdot \frac{n}{n+1}$$

$$= \underset{n\to\infty}{\text{lt.}} -\frac{\log n}{n} \cdot \frac{n}{1+\frac{1}{n}}$$

$$= 0 < 1$$

Hence by special logarithmic test, the given series is divergent when $x = \frac{1}{e}$.

Ex. 2. *Test the series*

$$1^p + \left(\frac{1}{2}\right)^p + \left(\frac{1.3}{2.4}\right)^p + \left(\frac{1.3.5}{2.4.6}\right)^p + \ldots$$

Sol. Leaving the first term, let the given series be denoted by Σu_n. Then,

$$u_n = \left(\frac{1.3.5.......2n-1}{2.4.6.....2n}\right)^p$$

$$\therefore \quad u_{n+1} = \left[\frac{1.3.5...(2n-1)(2n+1)}{2.4.6...(2n)(2n+2)}\right]^p$$

$$\frac{u_n}{u_{n+1}} = \left[\frac{2n+2}{2n+1}\right]^p$$

$$= \left[\frac{1+\dfrac{1}{n}}{1+\dfrac{1}{2n}}\right]^p$$

$$\therefore \quad \underset{n\to\infty}{\text{lt.}} \frac{u_n}{u_{n+1}} = 1$$

Hence ratio test fails.

Now apply logarithmic test,

we have

$$\log \frac{u_n}{u_{n+1}} = p\left[\log\left(1+\frac{1}{n}\right) - \log\left(1+\frac{1}{2n}\right)\right]$$

$$= p\left[\left(\frac{1}{n} - \frac{1}{2n^2} + \frac{1}{3n^3} \ldots\right) - \left(\frac{1}{2n} - \frac{1}{8n^2} + \frac{1}{24n^3} - \ldots\right)\right]$$

$$= p\left[\frac{1}{2n} - \frac{3}{8n^2} + \frac{7}{24n^3} - \ldots\right]$$

$$\therefore n \log \frac{u_n}{u_{n+1}} = p\left[\frac{1}{2} - \frac{3}{8n} + \frac{7}{24n^2} - \ldots\right]$$

$$\therefore \quad \underset{n\to\infty}{\text{lt.}} \; n\log\frac{u_n}{u_{n+1}} = \frac{p}{2}$$

Hence by logarithmic test, the given series is

convergent if $\frac{p}{2} > 1$ i.e. $p > 2$

divergent if $\frac{p}{2} < 1$ i.e. $p < 2$

and the test fails if $\frac{p}{2} = 1$ i.e. $p = 2$.

When $p = 2$,

apply special logarithmic test.

We have

$$n \log \frac{u_n}{u_{n+1}} = 1 - \frac{3}{4n} + \frac{7}{12n^2} - \ldots.$$

$$\therefore \quad n \log \frac{u_n}{u_{n+1}} - 1 = -\frac{3}{4n} + \frac{7}{12n^2} - \ldots.$$

$$\therefore \quad \log n \left[n \log \frac{u_n}{u_{n+1}} - 1 \right] - \frac{3}{4} \frac{\log n}{n} + \frac{7}{12} \frac{\log n}{n} \frac{1}{n} - \ldots$$

$$\therefore \quad \underset{n \to \infty}{\text{lt.}} \log n \left[n \log \frac{u_n}{u_{n+1}} - 1 \right] = 0 < 1$$

Hence the given series is divergent when $p = 2$.

Ex. 3. *Test the series*

$$1 + \left(\frac{1}{2}\right)^3 + \left[\frac{1.3}{2.4}\right]^3 + \left[\frac{1.3.5}{2.4.6}\right]^3 + \ldots$$

Sol. Proceed exactly as in Ex. 2 (p=3), the given series is convergent.

Ex. 4. *Test the series*

$$\left(\frac{1}{2 \cdot 4}\right)^p + \left(\frac{1 \cdot 3}{2 \cdot 4 \cdot 6}\right)^p + \left(\frac{1 \cdot 3 \cdot 5}{2 \cdot 4 \cdot 6 \cdot 8}\right)^p + \ldots$$

Sol. Let the given series be denoted by Σu_n. Then

$$u_n = \left[\frac{1 \cdot 3 \cdot 5 \ldots (2n-1)}{2 \cdot 4 \cdot 6 \ldots 2n(2n+2)} \right]^p$$

$$\therefore \quad u_{n+1} = \left[\frac{1 \cdot 3 \cdot 5 \ldots (2n-1)(2n+1)}{2 \cdot 4 \cdot 6 \ldots 2n(2n+2)(2n+4)} \right]^p$$

$$\therefore \quad \frac{u_n}{u_{n+1}} = \left(\frac{2n+4}{2n+1} \right)^p$$

$$= \left[\frac{2 + \frac{4}{n}}{2 + \frac{1}{n}} \right]^p$$

$$\therefore \quad \underset{n \to \infty}{\text{lt.}} \frac{u_n}{u_{n+1}} = 1$$

Hence ratio test fails.

Now applying logarithmic test,

we have,

$$\frac{u_n}{u_{n+1}} = \left(\frac{2n+4}{2n+1}\right)^p$$

$$= \left(1 + \frac{3}{2n+1}\right)^p$$

$$\therefore \quad \log \frac{u_n}{u_{n+1}} = p \log \left(1 + \frac{3}{2n+1}\right)$$

$$= p\left[\frac{3}{2n+1} - \frac{9}{2(2n+1)^2} + \ldots\right]$$

$$n \log \frac{u_n}{u_{n+1}} = p\left[\frac{3}{2n+1} - \frac{9}{2(2n+1)^2} + \ldots\right]$$

$$\therefore \quad \underset{n\to\infty}{\text{lt.}} \; n \log \frac{u_n}{u_{n+1}} = p \cdot \frac{3}{2}$$

Hence the given series is

convergent if $\frac{3p}{2} > 1$ i.e. $p > \frac{2}{3}$

divergent if $\frac{3p}{2} < 1$ i.e. $p < \frac{2}{3}$

and the test fails if $\frac{3p}{2} = 1$ i.e. $p = \frac{2}{3}$.

When $\quad p = \frac{2}{3}$,

apply special logarithmic test,

we have,

$$n \log \frac{u_n}{u_{n+1}} = \frac{2n}{2n+1} - \frac{3n}{(2n+1)^2} + \ldots$$

$$\therefore \quad n \log \frac{u_n}{u_{n+1}} - 1 = \frac{2n}{2n+1} - \frac{3n}{(2n+1)^2} + \ldots - 1$$

$$= \frac{-1}{2n+1} - \frac{3n}{(2n+1)^2} + \ldots$$

$$= \left[-\frac{n}{2n+1} - \frac{3n^2}{(2n+1)^2} + \ldots \right] \frac{1}{n}$$

$$\therefore \quad \log n \left[n \log \frac{u_n}{u_{n+1}} - 1 \right]$$

$$= \left[-\frac{n}{2n+1} - \frac{3n^2}{(2n+1)^2} \ldots\ldots \right] \frac{\log n}{n}$$

$$\therefore \quad \underset{n \to \infty}{\text{lt.}} \log n \left[n \log \frac{u_n}{u_{n+1}} - 1 \right] = 0 < 1$$

Hence the given series is divergent when $p = \frac{2}{3}$.

EXERCISE

Test the following series for convergence.

1. $1 + \frac{\alpha\beta}{1 \cdot \gamma} x + \frac{\alpha(\alpha+1)(\beta)(\beta+1)}{1 \cdot 2 \cdot \gamma \cdot (\gamma+1)} x^2$

$$+ \frac{\alpha(\alpha+1)(\alpha+2)(\alpha+3)\beta(\beta+1)(\beta+2)}{1 \cdot 2 \cdot 3 \cdot \gamma(\gamma+1)(\gamma+2)} x^3 + \ldots$$

Ans. Convergent when $x < 1$, divergent when $x > 1$. If $x = 1$, the series is convergent if $\gamma > \alpha + \beta$ and divergent if $\gamma \leq \alpha + \beta$.

2. $\Sigma = \frac{\alpha(\alpha+1)^2(\alpha+2)^2 + \ldots (\alpha+n-1)^2}{1 \cdot 2 \cdot 3 \ldots n \cdot \beta \cdot (\beta+1)(\beta+2) \ldots (\beta+n-1)} x^n$

Ans. Convergent when $x < 1$,
Divergent when $x > 1$,
If $x = 1$, convergent when $\beta > 2\alpha$ and divergent if $\beta \leq 2\alpha$.

3. If $\frac{u_n}{u_{n+1}} = \frac{n^k + An^{k-1} + Bn^{k-2} + Cn^{k-3} + \ldots}{n^k + an^{k-1} + bn^{k-2} + cn^{k-3} + \ldots}$

Where k is a +ve integer, show that the series Σu_n is convergent if $A - a - 1 > 0$ and divergent if $A - a - 1 \leq 0$.

2

Convergence and Divergence of Infinite Products

2.1 Definition

The product of an infinite no. of factors formed in given order according to some definite law is called an infinite product.

Let u_n be any real function of n defined for all +*ve* integral n then the product $u_1\, u_2\, u_3 \dots u_n$ is represented by P_n or by

$$\prod_{r=1}^{n} u_r$$

If the no. of factors in the product is infinite, then it is denoted by $\prod_{r=1}^{\infty} u_r$.

2.2 Convergent Product

If P_n-tends to a finite limit P other than zero, as n tends to infinity, the infinite product $\prod_{r=1}^{\infty} u_r$ converges to the limit P and we write

$$\prod_{r1}^{\infty} u_r = \lim_{n\to\infty} P_n = P$$

Note : None of the factors of P_n is zero for otherwise the infinite product redues to an insignificant number.

2.3 Divergent Product

If P_n does not tend to a finite non-zero limit then the infinite product said to diverge. If P_n tends to $+\infty$ or $-\infty$, the infinite product is said to diverge to $+\infty$ or $-\infty$, respectively and if P_n tends to zero when no factor of P_n is zero, then the infinite product is said to diverge to zero.

2.4 For the sake of convenience, the infinite product is generally written as

$$P_n = (1+u_1)(1+u_2)...(1+u_n)$$

2.5 To determine whether a given infinite product is finite or not.

Let the given infinite product be denoted by

$$P = u_1\, u_2\, u_3 ... u_n ...$$

Take logarithmic of both sides, we have,

$$\log P = \log u_1 + \log u_2 + \log u_3 + ... + \log u_n + ...$$
$$= \Sigma \log u_n$$
$$= \Sigma v_n \text{ (say) where } v_n = \log u_n$$

Now if Σv_n is convergent then $\log P$ is finite and if Σv_n is divergent then $\log P$ is infinite. Hence we reach the conclusion that,

$\log P$ is finite if Σv_n is convergent i.e. P is infinite when Σv_n is convergent and $\log P$ is infinite when Σv_n is divergent i.e.

$$\log P = +\infty \text{ or } -\infty \text{ or } P = e^{\infty} \text{ or } e^{-\infty} \text{ or } P = \infty \text{ or } 0$$

if Σv_n is divergent.

Illustrative Examples

Ex. 1. *Show that the following product is convergent*

$$\left(1+\frac{1}{1^2}\right)\left(1+\frac{1}{2^2}\right)\left(1+\frac{1}{3^2}\right)...\left(1+\frac{1}{n^2}\right)$$

Sol. Let $P = \left(1+\frac{1}{1^2}\right)\left(1+\frac{1}{2^2}\right)\left(1+\frac{1}{3^2}\right)...\left(1+\frac{1}{n^2}\right)$

$$\therefore \quad \log P = \log\left(1+\frac{1}{1^2}\right) + \log\left(1+\frac{1}{2^2}\right) + \log\left(1+\frac{1}{3^2}\right)$$
$$+ ... + \log\left(1+\frac{1}{n^2}\right)$$

$$= \Sigma \log\left(1+\frac{1}{n^2}\right) = \Sigma u_n \text{ (say)}$$

then $u_n = \log\left(1+\dfrac{1}{n^2}\right)$

$$= \frac{1}{n^2} - \frac{1}{2n^4} + \ldots$$

$$v_n = \frac{1}{n^2}$$

where v_n is the nth term of the auxiliary series Σv_n.

$$\frac{u_n}{v_n} = 1 - \frac{1}{2n^2} + \ldots$$

$$\therefore \quad \lim_{n\to\infty} \frac{u_n}{v_n} = 1$$

which is a fixed finite non-zero quantity.

Hence by comparison test, Σu_n and Σv_n converge or diverge together.

But $\Sigma v_n = \Sigma \dfrac{1}{n^2}$ is convergent as $p = 2(>1)$ by p-test

$\therefore$ Σu_n is also convergent by comparison test

$\therefore$ $\log P$ is finite

$\therefore$ P is finite.

Ex. 2. *Show that the infinite Product*

$$\left(1+\frac{1}{\sqrt{1}}\right)\left(1+\frac{1}{\sqrt{2}}\right)\left(1+\frac{1}{\sqrt{3}}\right)\ldots \text{diverges}$$

Sol. Let $P = \left(1+\dfrac{1}{\sqrt{1}}\right)\left(1+\dfrac{1}{\sqrt{2}}\right)\left(1+\dfrac{1}{\sqrt{3}}\right)\ldots$

$$\therefore \quad \log P = \log\left(1+\frac{1}{\sqrt{1}}\right) + \log\left(1+\frac{1}{\sqrt{2}}\right) + \log\left(1+\frac{1}{\sqrt{3}}\right) + \ldots$$

$$= \Sigma \log\left(1+\frac{1}{\sqrt{n}}\right)$$

$$= \Sigma u_n \text{ (say)}$$

$$\therefore \qquad u_n = \log\left(1+\frac{1}{\sqrt{n}}\right)$$

$$= \frac{1}{\sqrt{n}} - \frac{1}{2n} + \ldots$$

$$v_n = \frac{1}{\sqrt{n}}$$

where v_n is the nth term of the auxiliary series Σv_n.

$$\frac{u_n}{v_n} = 1 - \frac{1}{2n^{1/2}} + \ldots$$

$$\therefore \qquad \underset{n\to\infty}{\text{lt.}} \frac{u_n}{v_n} = 1$$

which is a fixed finite non-zero quantity.

Hence by comparison test, Σu_n and Σv_n converge or diverge together.

But $\Sigma v_n = \Sigma \frac{1}{n^{1/2}}$ is divergent as $P\left(=\frac{1}{2}\right)<1$ by p-test

Hence Σu_n is also divergent, by comparison test.

$\therefore$ $\log P$ is infinite when $n \to \infty$

$\therefore$ P is infinite when $n \to \infty$.

Ex. 3. *Show that the infinite product*

$$\left(1-\frac{1}{2^2}\right)\left(1-\frac{1}{3^2}\right)\left(1-\frac{1}{4^2}\right)\ldots \text{ is convergent}$$

Sol. Proceed exactly as in Ex. 1.

Ex. 4. *Show that the product*

$$\left(1+\frac{1}{1^3}\right)\left(1+\frac{1}{2^3}\right)\left(1+\frac{1}{3^3}\right)\ldots\left(1+\frac{1}{n^3}\right)$$

is finite when $n \to \infty$.

Sol. Proceed exactly as in Ex. 1.

Ex. 5. *Show that the product*

$$\left(1+\frac{1}{1^{3/2}}\right)\left(1+\frac{1}{2^{3/2}}\right)...\left(1+\frac{1}{n^{3/2}}\right)$$

is convergent.

Sol. Proceed exactly as in Ex. 1.

Ex. 6. *Show that the product*

$$\left(1+\frac{1}{2}\right)\left(1+\frac{1}{3}\right)\left(1+\frac{1}{4}\right)...$$

is divergent.

Sol. Proceed exactly as in Ex. 2.

Ex. 7. *Show that the product*

$$\frac{1}{2}.\frac{3}{2}.\frac{3}{4}.\frac{5}{4}.\frac{5}{6}.\frac{7}{6}...\frac{2n-1}{2n}.\frac{2n+1}{2n}$$

is finite when n is infinite.

Sol. Let $P=\left(\frac{1}{2}.\frac{3}{2}\right)\left(\frac{3}{4}.\frac{5}{4}\right)\left(\frac{5}{6}.\frac{7}{6}\right)...\left(\frac{2n-1}{2n}.\frac{2n+1}{2n}\right)$

$$\therefore \quad \log P=\log\left(\frac{1}{2}\cdot\frac{3}{2}\right)+\log\left(\frac{3}{4}\cdot\frac{5}{4}\right)+\log\left(\frac{5}{6}\cdot\frac{7}{6}\right)$$

$$+...+\log\left(\frac{2n-1}{2n}.\frac{2n+1}{2n}\right)$$

$$=\Sigma\log\left(\frac{2n-1}{2n}.\frac{2n+1}{2n}\right)=\Sigma u_n \text{ (say)}$$

$$\therefore \quad u_n=\log\left(\frac{2n-1}{2n}.\frac{2n+1}{2n}\right)$$

$$=\log\left(\frac{4n^2-1}{4n^2}\right)=\log\left[1-\frac{1}{4n^2}\right]$$

$$=-\left[\frac{1}{4n^2}+\frac{1}{2.(4n^2)^2}+\frac{1}{3}\frac{1}{(4n^2)^3}+...\right]$$

$$= -\left[\frac{1}{4n^2} + \frac{1}{32n^4} + \frac{1}{192n^6} + \dots\right]$$

$$v_n = \frac{1}{n^2}$$

where v_n is n^{th} term of the auxiliary series Σv_n.

$$\therefore \quad \frac{u_n}{v_n} = -\left[\frac{1}{4} + \frac{1}{32n^2} + \frac{1}{192n^4} + \dots\right]$$

$$\therefore \quad \underset{n \to \infty}{\text{lt.}} \; -\frac{u_n}{v_n} = -\frac{1}{4}$$

which is a fixed finite non-zero quantity.

Hence by comparison test, Σu_n and Σv_n both converge or diverge.

But $\Sigma v_n = \Sigma \frac{1}{n^2}$ is convergent as $p = 2(>1)$ by *p*-test.

Hence Σu_n is also convergent by comparison test i.e. the sum of the series Σu_n is finite as $n \to \infty$ i.e. log *P* is finite when $n \to \infty$ i.e. *P* is finite when *n* is infinite.

Ex. 8. *Show thdt the product*

$$\frac{2}{1}.\frac{2}{3}.\frac{4}{3}.\frac{4}{5}.\frac{6}{5}.\frac{6}{7}\dots.\frac{2n-2}{2n-3}\,\frac{2n-2}{2n-1}.\frac{2n}{2n-1}$$

is finite when n tend to infinity.

Sol. Let $P = \left(\frac{2}{1}.\frac{2}{3}\right)\left(\frac{4}{3}.\frac{4}{5}\right)\left(\frac{6}{5}.\frac{6}{7}\right)\dots\left(\frac{2n-2}{2n-3}.\frac{2n-2}{2n-1}\right)\frac{2n}{2n-1}$

$$\therefore \quad \log P = \left(\frac{2}{1}\cdot\frac{2}{3}\right) + \log\left(\frac{4}{3}\cdot\frac{4}{5}\right) + \log\left(\frac{6}{5}\cdot\frac{6}{7}\right)$$

$$+ \dots \log\left(\frac{2n-2}{2n-3}\cdot\frac{2n-2}{2n-1}\right) + \log\left(\frac{2n}{2n-1}\right)$$

$$= \Sigma \log \frac{(2n-2)(2n-2)}{(2n-3)(2n-1)} + \log\left(\frac{2n}{2n-1}\right)$$

Let $$\Sigma u_n = \Sigma \log \frac{(2n-2)(2n-2)}{(2n-3)(2n-1)}$$

Then $$u_n = \log \frac{(2n-2)(2n-2)}{(2n-3)(2n-1)}$$

or, $$u_n = \log \frac{4n^2 - 8n + 4}{4n^2 - 8n + 3}$$

$$= \log \left(1 + \frac{1}{4n^2 - 8n + 3}\right)$$

$$= \frac{1}{4n^2 - 8n + 3} - \frac{1}{2(4n^2 - 8n + 3)^2} + \ldots$$

$v_n = \frac{1}{n^2}$ where $\Sigma v_n = \Sigma \frac{1}{n^2}$ is the auxiliary series.

Then, $$\frac{u_n}{v_n} = \frac{1}{4 - \frac{8}{n} + \frac{3}{n^2}} - \frac{1}{2\left(4n - 8 + \frac{3}{n}\right)^2} + \ldots$$

$$\therefore \quad \underset{n\to\infty}{\text{lt.}} \frac{u_n}{v_n} = \frac{1}{4}$$

which is a fixed finite non-zero quantity.

Hence by comparison test, Σu_n and Σv_n converge or diverge together.

But $\Sigma \frac{1}{n^2}$ is convergent as $p = 2 (> 1)$ by p-test.

Hence Σu_n is also convergent i.e. the sum of the series Σu_n is finite when n is infinite.

Also $$\underset{n\to\infty}{\text{lt.}} \log \frac{2n}{2n-1}$$

$$= \underset{n\to\infty}{\text{lt.}} \log \left[\frac{1}{1 - \frac{1}{2n}}\right]$$

$$= \underset{n \to \infty}{\text{lt.}} \left[\log 1 - \log\left(1 - \frac{1}{2n}\right)\right]$$

$$= \underset{n \to \infty}{\text{lt.}} \ -\log\left(1 - \frac{1}{2n}\right)$$

$$= \underset{n \to \infty}{\text{lt.}} \left[\frac{1}{2n} + \frac{1}{2}\frac{1}{(2n)^2} + \ldots\right]$$

$$= 0$$

$\therefore$ $\log P$ is convergent when $n \to \infty$

i.e. P is finite when $n \to \infty$.

Ex. 9. *Show that the product*

$$\left(1 - \frac{1}{2^2}\right)\left(1 - \frac{1}{3^2}\right)\left(1 - \frac{1}{4^2}\right)\ldots$$

converges to $\frac{1}{2}$.

Sol. Here,

$$P_{n-1} = \left(1 - \frac{1}{2^2}\right)\left(1 - \frac{1}{3^2}\right)\left(1 - \frac{1}{4^2}\right)\ldots\left(1 - \frac{1}{n^2}\right)$$

$$= u_1 \, u_2 \, u_3 \ldots u_{n-1}$$

where $u_{n-1} = 1 - \frac{1}{n^2}$

$$= \frac{n^2 - 1}{n^2}$$

$$= \frac{(n-1)(n+1)}{n^2} \qquad \ldots(1)$$

Put $n = 2, 3, \ldots$ etc. in (1), we have

$$u_1 = \frac{1.3}{2^2},$$

$$u_2 = \frac{2.4}{3^2}$$

$$u_3 = \frac{3.5}{4^2}$$

$$u_4 = \frac{4.6}{5^2}$$

Multiplying all the factors, we have

$$P_{n-1} = \left(\frac{1}{2}\cdot\frac{3}{2}\right)\left(\frac{2}{3}\cdot\frac{4}{3}\right)\left(\frac{3}{4}\cdot\frac{5}{4}\right)\left(\frac{4}{5}\cdot\frac{6}{5}\right)\cdots$$

$$\left(\frac{n-2}{n-1}\cdot\frac{n}{n-1}\right)\left(\frac{n-1}{n}\cdot\frac{n+1}{n}\right)$$

$$= \frac{1}{2}\cdot\frac{n+1}{n}$$

$$= \frac{1}{2}\left(1+\frac{1}{n}\right)$$

$$\therefore \quad \underset{n\to\infty}{\text{lt.}} P_{n-1} = \frac{1}{2}$$

Also $$P_n = P_{n-1}\left[1-\frac{1}{(n+1)^2}\right]$$

$$= P_{n+1}\frac{(n+1)^2-1^2}{(n+1)^2}$$

$$= P_{n-1}\frac{(n+1-1)(n+1+1)}{(n+1)^2}$$

$$= P_{n-1}\frac{n(n+2)}{(n+1)^2}$$

$$= P_{n-1}\frac{\left[1+\frac{2}{n}\right]}{\left[1+\frac{1}{n}\right]^2}$$

$$\therefore \quad \underset{n\to\infty}{\text{lt.}} \; P_n = \underset{n\to\infty}{\text{lt.}} \; P_{n-1} \frac{\left[1+\frac{2}{n}\right]}{\left[1+\frac{1}{n}\right]^2} = \frac{1}{2}$$

Ex. 10. *Show that the product*

$$\left(1+\frac{1}{2}\right)\left(1-\frac{1}{3}\right)\left(1+\frac{1}{4}\right)\left(1-\frac{1}{5}\right) \ldots \textit{ converges to } 1$$

Sol. $P_{2n} = \left(1+\frac{1}{2}\right)\left(1-\frac{1}{3}\right)\left(1+\frac{1}{4}\right)\left(1-\frac{1}{5}\right) \ldots + \left(1+\frac{1}{2n}\right)\left(1-\frac{1}{2n+1}\right)$

$$= \frac{3}{2}.\frac{2}{3}.\frac{5}{4}.\frac{4}{5} \ldots \frac{2n+1}{2n}.\frac{2n}{2n+1}$$

$$= 1$$

Also, $\quad P_{2n+1} = P_{2n}\left[1+\frac{1}{2n+2}\right]$

$$\therefore \quad \underset{n\to\infty}{\text{lt.}} \; P_{2n+1} = \underset{n\to\infty}{\text{lt.}} \left[P_{2n}\left[1+\frac{1}{2n+2}\right]\right]$$

$$= \underset{n\to\infty}{\text{lt.}} \; P_{2n}$$

$$= 1$$

Hence $P_n \to 1$ when $n \to \infty$.

whether n is even or odd.

Therefore the given product converges to 1.

Ex. 11. *Show that the infinite product*

$$\left(1+\frac{x}{1^2}\right)\left(1+\frac{x}{2^2}\right)\left(1+\frac{x}{3^2}\right)\ldots$$

is convergent for all values of x.

Sol.

Let $\quad P = \left(1+\frac{x}{1^2}\right)\left(1+\frac{x}{2^2}\right)\left(1+\frac{x}{3^2}\right)\ldots\left(1+\frac{x}{n^2}\right)\ldots$

$$\log P = \log\left(1+\frac{x}{1^2}\right)+\log\left(1+\frac{x}{2^2}\right)$$
$$+\log\left(1+\frac{x}{3^2}\right)+\ldots+\log\left(1+\frac{x}{n^2}\right)+\ldots$$

$$= \Sigma \log\left[1+\frac{x}{n^2}\right] = \Sigma u_n \text{ (say)}$$

$$\therefore \quad u_n = \log\left(1+\frac{x}{n^2}\right)$$

$$= \frac{x}{n^2} - \frac{1}{2}\frac{x^2}{n^4} + \ldots$$

$$v_n = \frac{1}{n^2}$$

where v_n is n^{th} term of the auxiliary series Σv_n.

$$\frac{u_n}{v_n} = x - \frac{1}{2}\frac{x^2}{n^2} + \ldots.$$

$$\therefore \quad \lim_{n\to\infty} \frac{u_n}{v_n} = x$$

which is a finite quantity.

Hence by comparison test, Σu_n and Σv_n converge or diverge together.

But $\Sigma v_n = \Sigma \frac{1}{n^2}$ is convergent as $p = 2(>1)$ by p-test

Hence Σu_n is also convergent.

$\therefore$ $\log P$ is convergent.

Hence P i.e. the given infinite product is also convergent for all values of x.